SpringerBriefs on Case Studies of Sustainable Development

Series editors

Asit K. Biswas, Third World Centre for Water Management, Los Clubes, Atizapan, Mexico
Cecilia Tortajada, Los Clubes, Atizapan, Mexico

The importance of sustainable development has been realized for at least 60 years, even though the vast majority of people erroneously think this concept originated with the Brundtland Commission report of 1987 on Our Common Future. In spite of at least six decades of existence, we only have some idea as to what is NOT sustainable development rather than what is. SpringerBriefs on Case Studies of Sustainable Development identify outstanding cases of truly successful sustainable development from different parts of the world and analyze enabling environments in depth to understand why they became so successful. The case studies will come from the works of public sector, private sector and/or civil society. These analyses could be used in other parts of the world with appropriate modifications to account for different prevailing conditions, as well as text books in universities for graduate courses on this topic. The series of short monographs focuses on case studies of sustainable development bridging between environmental responsibility, social cohesion, and economic efficiency. Featuring compact volumes of 50 to 125 pages (approx. 20,000–70,000 words), the series covers a wide range of content—from professional to academic—related to sustainable development. Members of the Editorial Advisory Board: Mark Kramer, Founder and Managing Director, FSG, Boston, MA, USA Bernard Yeung, Dean, NUS Business School, Singapore.

More information about this series at http://www.springer.com/series/11889

Nur Sabahiah Abdul Sukor
Nur Khairiyah Basri

Travel Behaviour Modification (TBM) Program for Adolescents in Penang Island

Intervention Ideas to Promote Sustainable Transport

 Springer

Nur Sabahiah Abdul Sukor
School of Civil Engineering
Universiti Sains Malaysia
Nibong Tebal, Penang
Malaysia

Nur Khairiyah Basri
School of Civil Engineering
Universiti Sains Malaysia
Nibong Tebal, Penang
Malaysia

ISSN 2196-7830   ISSN 2196-7849   (electronic)
SpringerBriefs on Case Studies of Sustainable Development
ISBN 978-981-13-2504-5   ISBN 978-981-13-2505-2   (eBook)
https://doi.org/10.1007/978-981-13-2505-2

Library of Congress Control Number: 2018955158

This Springer imprint is published by the registered company Springer Nature Singapore Pte Ltd.
The registered company address is: 152 Beach Road, #21-01/04 Gateway East, Singapore 189721, Singapore

We dedicate this book to those who young (at heart)...

Preface

This book presents the case study of travel behaviour modification (TBM) techniques that focused on the adolescents in Penang Island. The earlier section of the book deliberates at length the importance of sustainable transport and Sustainable Development Goals (SDGs) agendas. Readers will also be acquainted with examples of adolescents' daily travel patterns during weekends and weekdays that included the simple and complex travel patterns.

The book continues with the overview of travel behaviour modification (TBM) strategies that had been experimented in the developed countries such as Australia, Japan and Europe. The important elements, which are the interventions that had been done for travel behaviour modification (TBM) in Penang Island, were provided in this book. The flow of the programme was explained in detail. This book also provides the consequences and the implication of the travel behaviour modification (TBM) to the adolescents' daily travel patterns in Penang Island such as choice of the transportation, trip purpose, travel time, travel distance and CO_2 emissions.

Therefore, researchers, academicians and students will find this book useful to enhance their comprehension on the subject matter, particularly to those who are interested in encouraging sustainable transport and travel modification techniques in marketing strategies. Likewise, this book will also be helpful for city planners and public transport operators in promoting public transport services in their city, especially bus transit services.

The authors foremost wish to acknowledge the financial support given by Universiti Sains Malaysia (USM) through the APEX grant (1002.PAWAM.910346) and Ministry of Higher Education Malaysia through Trans-disciplinary Research Grant Scheme (TRGS) (203.PAWAM.67610002) that enabled this book to be written. We also take this opportunity to express our deepest gratitude to Rapid Penang, selected schools involved in the case study and volunteers for their strong support and commitment.

Nibong Tebal, Malaysia   Nur Sabahiah Abdul Sukor
July 2018   Nur Khairiyah Basri

Contents

About the Authors

Nur Sabahiah Abdul Sukor is a senior lecturer and researcher in the School of Civil Engineering, Universiti Sains Malaysia. She received her Ph.D. in transportation engineering from Kyoto University in 2011. Currently, her areas of specializations are sustainable transport and travel behaviour where most of her publications emphasize the outcomes for Sustainable Development Goals, especially SDG 11.

Nur Khairiyah Basri gained her Ph.D. in sustainable transportation from School of Civil Engineering, Universiti Sains Malaysia. During her Ph.D. programme, she worked as a researcher and teaching assistant in Dr. Sukor's research group. Her tasks were to design and analyse the survey and interviews focusing the secondary school students. Her research interest is statistical modelling, especially that relates to transportation.

List of Figures

List of Tables

Chapter 1
Sustainable Transports

Abstract This chapter elaborates the perspective of sustainable transports by explaining the benefits of using public transports and active modes. As well as to reduce the number of private vehicles on the roads, sustainable transports also aimed to reduce the road accidents. In addition, sustainable transport is among the big agendas in Sustainable Development Goals (SDGs) that need to be taken into account for planning future cities for our generation. Together with the explanation on sustainable transports, this chapter also highlights several improvements that can be executed in the planning for a sustainable transportation system by adapting the avoid-promote-alternative mitigation strategies.

Keywords Sustainable transport · Sustainable development goal (SDG) Public transport · Low carbon · Mode shift

1.1 Perspectives on Sustainable Transports

With cars as the primary form of transport for daily travel in the developing countries within the South East Asia region, the road traffic is undeniably dreadful, especially during the rush hour, holidays and festive season. Furthermore, the bus services are not particularly favoured due to the longer travel duration compared to travelling by cars. Besides that, motorcycles, as another form of transport, offer another alternative in beating the road traffic, but this alternative comes with higher risk of road accident and in the worst case, death. The cars and motorcycles are deemed as unsustainable means of transports, but to each his own in selecting the preferred form of transport to travel from one destination to another.

Imagining the need of not having to go through the hectic road traffic during the rush hour just for lunch—truly, it is a pleasure and stress reliever to enjoy the greens along the way back to the office, which is merely a walking distance away. Without a doubt, such an experience is possible in certain cities across the globe, but rather uncommon in most South East Asian countries. Meanwhile, the communities in developed countries, such as Japan and Europe, demonstrate the

© The Author(s) 2019
N. S. A. Sukor and N. K. Basri, *Travel Behaviour Modification (TBM) Program for Adolescents in Penang Island*, SpringerBriefs on Case Studies of Sustainable Development, https://doi.org/10.1007/978-981-13-2505-2_1

propensity of selecting bicycles as their preferred mode of transport for short-distance travel.

With the convenience of cycling tracks, the need for private vehicles to move about is lower. The cycling tracks are well-connected and the cyclists do not share these tracks with drivers of other forms of transport. However, there are certain drawbacks, particularly when it comes to wet season or winter. Besides that, cycling uphill on a steep road with heavyweight load in summer may be a tremendous feat for most. Nevertheless, cycling through the summer breeze under the tree shades can be a real delight. When it comes to the weekends, most people are likely to opt for public transports, such as the bus or train, for the long-distance travel. The commuters have the opportunity to make the most of their commute, especially during the transits, with light activities, such as reading, listening to music or even grab a bite along the way.

The commuting experience affects the public choice of transport; thus, the monotonous commute can be a put-off. The pleasantness of the public space is often identified as one of the factors that prompt the public to opt for public transports, cycling and even walking (Bissell 2010; Myers 2011). However, the question remains—why shifting the mode from private motor vehicles to active mode or public transport does important? Is it merely to beat the road traffic or is there a more important agenda here? What do we know about the sustainable transports and why is this topic significant? The subsequent sub-chapter provides insights to unravel these questions.

1.2 Definition of Sustainable Transports

There are various definitions of 'sustainable transport'. Notably, sustainable transport addresses the present transport needs without affecting the transport needs of future generations (Black 1996; Richardson 2005), which is in line with one of the Sustainable Development Goals (SDGs). Besides that, the concept of sustainable transports propounds low-carbon and energy-efficient transports in the urban settings (Awasthi et al. 2011; Lu and Pas 1999). Public transports, such as bus, mass transit and non-motorized transports (such as cycling and walking) are considered as sustainable transports with lower carbon emissions (Limanond et al. 2011; Lah 2017).

Basically, sustainable transports have lower carbon emissions and energy consumption, because energy is also said to release emissions. Despite the advancement of green transportation technologies, the transportation industry remains as one of the primary sources of carbon emissions through the combustion of fossil fuel (Wang et al. 2014). Apart from the worsening congestion and air pollution, the heavy dependence on private vehicles to move about on the daily basis highlights the large uncertainty in the future of fossil fuel and the need for stricter carbon emissions standards (Gössling 2013; Sang and Bekhet 2015). Addressing that, the Fifth Assessment Report (AR5) of the Intergovernmental Panel on Climate Change

(IPCC) outlined three features in the mitigation strategies towards low carbon (Pachauri et al. 2015): (1) Avoid the use of private vehicles to travel (including the reduction in the travel distance) through the implementation of compact cities and multimodal communities; (2) Promote the use of public transports or non-motorized transports; (3) Alternative fuels and advanced transportation technologies for improved environmental performance. Figure 1.1 provides several strategies to improve the overall transportation system with respect to the sustainable development goals.

However, solely depending on technological advancement is not an ideal strategy to achieve low-carbon mobility (Givoni 2013). It is because of the mobility issues were more likely to be one of travel behaviour problems. Therefore, studies on travel behaviour changes are recommended in the realization of the sustainable low-carbon transport system, which can be implemented in varieties of directions (Taylor 2007; Line et al. 2012). Accordingly, the government should be at the forefront of the sustainability development by promoting the use of public transports (Wang and Chang 2014). Majority of the public transport commuters in most cities, particularly in the developing countries, are of those with lower income. In other words, they are pressed to experience longer travel duration. Public transport services should be revised for the use of commuters from all backgrounds at reasonable price and higher efficiency to attain the status of sustainable city. Despite

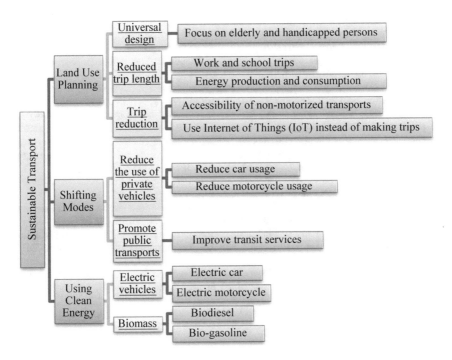

Fig. 1.1 Strategies for improved transport system towards sustainable development (*Source:* Shiau 2013)

the improvement in the quality of public transports, the public resistance in using the public transports is of another concern, which highlights the significance of travel behaviour modification (TBM). The following chapters explore the concept of TBM in detail.

References

Awasthi A, Chauhan SS, Omrani H (2011) Application of fuzzy TOPSIS in evaluating sustainable transportation systems. Expert Syst Appl 38:12270–12280. https://doi.org/10.1016/j.eswa. 2011.04.005

Bissell D (2010) Passenger mobilities: affective atmospheres and the sociality of public transport. Environ Plan D 28:270–289. https://doi.org/10.1068/d3909

Black WR (1996) Sustainable transportation: a US perspective. J Transp Geogr 4:151–159. https://doi.org/10.1016/0966-6923(96)00020-8

Givoni M (2013) Alternative pathways to low carbon mobility. In: Givoni M, Banister D (eds) Moving towards low carbon mobility, 1st edn. Edward Elgar Publishing, Cheltenham, UK

Gössling S (2013) Urban transport transitions: Copenhagen, city of cyclists. J Transp Geogr 33:196–206. https://doi.org/10.1016/j.jtrangeo.2013.10.013

Lah O (2017) Sustainable development synergies and their ability to create coalitions for low-carbon transport measures. Trans Res Proc 25:5088–5098. https://doi.org/10.1016/j.trpro. 2017.05.495

Limanond T, Butsingkorn T, Chermkhunthod C (2011) Travel behaviour of university students who live on campus: a case study of a rural university in asia. Transp Policy 18:163–171. https://doi.org/10.1016/j.tranpol.2010.07.006

Line T, Chatterjee K, Lyons G (2012) Applying behavioural theories to studying the in fluence of climate change on young people's future travel intentions. Transp Res D Transp Environ 17:270–276. https://doi.org/10.1016/j.trd.2011.12.004

Lu X, Pas EI (1999) Socio-demographics, activity participation and travel behavior. Transp Res Part A Policy Pract 33:1–18. https://doi.org/10.1016/S0965-8564(98)00020-2

Myers M (2011) Walking again lively: towards an ambulant and conversive methodology of performance and research. Mobilities-UK 6:183–201. https://doi.org/10.1080/17450101.2011.552775

Pachauri RK, Meyer L, Plattner GK, Stocker T (2015) Climate change 2014: synthesis report. Contribution of Working Groups I, II and III to the fifth assessment report of the Intergovernmental Panel on Climate Change, p 151. IPCC, Geneva, Switzerland

Richardson BC (2005) Sustainable transport: analysis frameworks. J Transp Geogr 13:29–39. https://doi.org/10.1016/j.jtrangeo.2004.11.005

Sang YN, Bekhet HA (2015) Modelling electric vehicle usage intentions: an empirical study in Malaysia. J Clean Prod 92:75–83. https://doi.org/10.1016/j.jclepro.2014.12.045

Shiau TA (2013) Evaluating sustainable transport strategies for the counties of Taiwan based on their degree of urbanization. Transp Policy 30:101–108. https://doi.org/10.1016/j.tranpol.2013. 09.001

Taylor MA (2007) Voluntary travel behavior change programs in Australia: the carrot rather than the stick in travel demand management. Int J Sustain Transp 1:173–192. https://doi.org/10. 1080/15568310601092005

Wang N, Chang YC (2014) The evolution of low-carbon development strategies in China. Energy 68:61–70. https://doi.org/10.1016/j.energy.2014.01.060

Wang J, Chi L, Hu X, Zhou H (2014) Urban traffic congestion pricing model with the consideration of carbon emissions cost. Sustainability 6:676–691. https://doi.org/10.3390/su6020676

Chapter 2
Adolescents as the Target Users for Sustainable Transports

Abstract It is important to understand the daily travel patterns of the adolescents as a strategy to nurture them to be more pro-sustainable transport-users in the future. Usually, the adolescents need to depend on their parents to travel either during weekdays or weekends. This chapter addressed why adolescents had been a target group for travel behaviour modification studies and what factors affecting their decision in making trips. From the survey that had been done to the adolescents in Penang Island, the finding shows that the adolescents were tended to involve with simple travel patterns (a single-trip chain in one trip) and complex travel patterns (many trip chains in one trip) after the school hours. The travel patterns were also found to be different between weekdays and weekends.

Keywords Adolescents · Travel behaviour · Travel pattern · Trip chain School trip · Non-school trip

2.1 Who Are the Adolescents?

The adage "as the twig is bent, so is the tree inclined" suggests the importance of inspiring and educating adolescents (of between 13 and 18 years of age)—an important, significant phase towards adulthood (Clifton 2003; Datz et al. 2005). It is essential that these young individuals are aware of the significance of sustainable transports, such as the use of public transports or the options of cycling or walking. Although their travel behaviour is yet to be moulded (Kamargianni 2010), certain studies found that the adolescents demonstrate the intention of owning private vehicles, which can be influenced by identity, image, social recognition of the transport services, the adolescents' attitude towards transport modes, (Turner and Pilling 1999; Storey and Brannen 2000; Line et al. 2010), and even their parents' attitude towards transport modes (Nishihara et al. 2017).

Besides that, it was also revealed that the adolescents have the propensity of making independent travel decisions following their transitional phase towards adulthood (Clifton 2003), but their travel behaviour is often overseen by their

© The Author(s) 2019

N. S. A. Sukor and N. K. Basri, *Travel Behaviour Modification (TBM) Program for Adolescents in Penang Island*, SpringerBriefs on Case Studies of Sustainable Development, https://doi.org/10.1007/978-981-13-2505-2_2

parents (Mackett 2013). Most parents are likely to depend on private vehicles to travel on daily basis due to safety and security concerns, complications in the daily family schedule, or travel distance issue (Mackett et al. 2007; Bringolf-Isler et al. 2008; Carver et al. 2010; McDonald et al. 2010; Nasrudin and Nor 2013).

The travel behaviours of working adults and adolescents (to schools) exhibit almost similar routine—daily commute with fixed time and location (Copperman 2008; Kamargianni 2014). The adolescents typically commute between 0700 and 0900 h, which is rather similar to the time working adults commute to work (De Dios Ortuzar and Willumsen 2001). Despite that, the activities these adolescents participate on daily basis are more dynamic in nature, which require different mobility decisions compared to the adults (Kamargianni 2014). For instance, they travel to attend extra academic class, sports and leisure activities, such as shopping or meeting up with friends (Bowman and Ben-Akiva 2001; Clifton 2003; Marzuki and Mohamad 2006). The parents' travel behaviour is affected by getting their children to their next destination is part of their daily schedule (Paleti et al. 2011) given their dependence on private vehicles. Therefore, there is a need for these adolescents to independently make their travel decisions. The subsequent sub-chapter presents the travel behavioural pattern of the adolescents in Penang Island.

2.2 Travel Behavioural Pattern of Adolescents in Penang Island

Located at the northern region of Malaysia, Penang Island occupies a land area of 285 km^2 within the state of Penang (1038 km^2). The traffic congestion is a prevalent issue in Penang Island, particularly during the rush hour or wet season, with the heavy dependence on private vehicles as the primary form of transport and a substantial number of public and private schools (589). The use of public transports, such as bus, among the adolescents, remains minimal. Therefore, it is crucial to grasp the travel behaviour of adolescents for the development of an efficient public transport system in terms of accessibility, facilities, infrastructure, and security to encourage the use of public transports. The assessment of travel behaviour of adolescents typically considers their trip-chaining pattern, the purpose of the travel, travel duration, travel distance and modal split.

The trip chaining involves a sequence of trips between the point of departure and the destination that includes the stops, while the purpose of travel influences the daily trip chaining. In the case of Penang Island, the trip-chaining pattern of adolescents during the weekdays and weekends according to the respective activities demonstrated eight primary characteristics. As shown in Fig. 2.1, the travels during the weekdays for the adolescents mainly consist of simple trip chain between home (denoted as H) and school (denoted as SC), which is presented as **H – SC – H**. These two destinations in terms of schedule and location are set for adolescents on

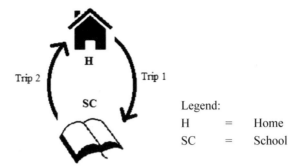

Fig. 2.1 Example of simple trip chain of H – SC – H on weekdays

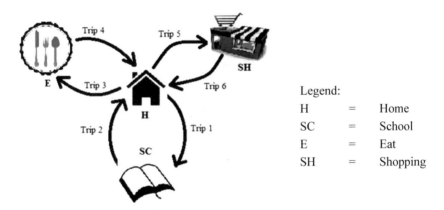

Fig. 2.2 Example of complex trip chain of H – SC – H – [+(NSC – H)] on weekdays

daily basis. In most cases, the adolescents are more likely to remain home after school.

Meanwhile, complex trip chaining involves several stops between first outdoor activity (away from home) and the final outdoor activity. In this case, it involves non-school activities (denoted as NSC), which include extra academic class, meeting up with relatives or friends, sports, leisure activities, dining out and/or other personal business activities. There are three types of travel behavioural pattern that demonstrate complex trip chaining among the adolescents during the weekdays. The first complex trip chain of **H – SC – H – [+(NSC – H)]** (Fig. 2.2) suggests that the adolescents make the usual trips between home and school and a sequence of trips between home and other destinations for non-school activities. Meanwhile, the second complex trip chain of **H – SC – H – (+NSC) – H** (Fig. 2.3) implies that after making the usual trips between home and school, the adolescents make a trip to participate in several non-school activities before returning home again. The final complex trip chain on weekdays, specifically **H – SC – (+NSC) –**

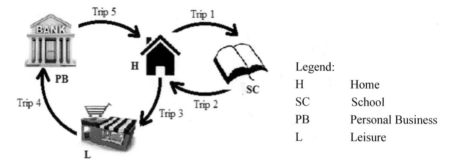

Fig. 2.3 Example of complex trip chain of H – SC – H – (+NSC) – H on weekdays

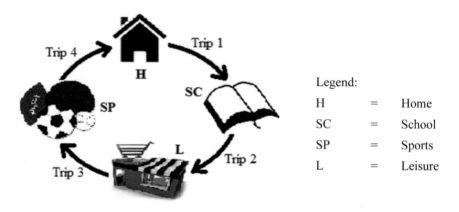

Fig. 2.4 Example of complex trip chain of H – SC – (+NSC) – H on weekdays

H (Fig. 2.4), differs from the above. In this case, the adolescents make the usual trip from home to school, but do not return home after school. Instead, they participate in the non-school activities before returning home.

On the other hand, the concept of the travel behavioural pattern on weekends is rather similar to the travel behavioural pattern on weekdays among the adolescents. The trips ensue from home and end at home with various stops that involve the destinations for only non-school activities. Apart from the simple trip chaining, there are three types of travel behavioural pattern that demonstrate complex trip chaining among the adolescents during the weekends.

The simple trip chain, denoted as **H – SNC – H** (Fig. 2.5), implies that the adolescents, in this case, travel between two destinations only for the day (during the weekends). In most cases, the adolescents are more likely to remain home after the specific non-school activity. The first identified complex trip chain for the weekends is denoted as **H – [+(NSC – H)]** (Fig. 2.6), which suggests that the adolescents make several trips between home and the destinations for non-school activities. Example of schedule that reflects this particular complex trip chaining is

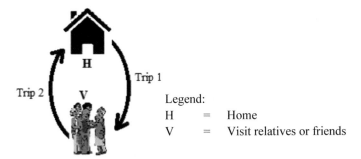

Fig. 2.5 Example of simple trip chain of H – NSC – H on weekends

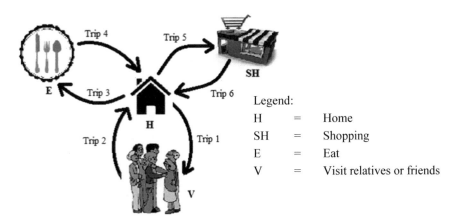

Fig. 2.6 Example of complex trip chain of H – [+(NSC – H)] on weekends

making a trip (from home) to meet up with friends or other family members (Activity 1) and return home before making another trip (from home) to dine out (Activity 2), but only to return home again before making another trip (from home) to shop (Activity 3) to end the travel for the day (return home).

The second identified complex trip chain for the weekends is presented as **H – NSC – H – (+NSC – H)** (Fig. 2.7). In this case, the adolescents make a trip between home and a particular destination for a specific non-school activity (e.g., breakfast). Following that, they make another trip (from home) with several stops to different destinations for various non-school activities prior to returning home. The final complex trip chain on weekends, specifically **H – (+NSC) – H** (Fig. 2.8), differs from the previous complex trip chains on weekends. In this case, the adolescents travel from home to the various destinations for non-school activities and only return home after completing the trips.

Effective strategies to encourage these adolescents to make use of sustainable transports, specifically the bus, for their daily travel can be appropriately planned

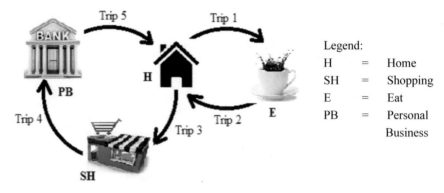

Fig. 2.7 Example of complex trip chain of H – NSC – H – (+NSC) – H on weekends

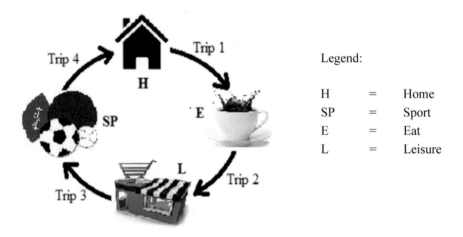

Fig. 2.8 Example of complex trip chain of H – (+NSC) – H on weekends

and implemented with the in-depth understanding of their travel behavioural pattern. The following chapter presents various intervention strategies to propel transformed travel behaviour towards sustainable transports.

References

Bowman JL, Ben-Akiva ME (2001) Activity-based disaggregate travel demand model system with activity schedules. Transp Res Rec 35:1–28. https://doi.org/10.1016/S0965-8564(99)00043-9

Bringolf-Isler B, Grize L, Mäder U, Ruch N, Sennhauser FH, Braun-Fahrländer C (2008) Personal and environmental factors associated with active commuting to school in Switzerland. Prev Med 46:67–73. https://doi.org/10.1016/j.ypmed.2007.06.015

Carver A, Timperio A, Hesketh K, Crawford D (2010) Are children and adolescents less active if parents restrict their physical activity and active transport due to perceived risk? Soc Sci Med 70:1799–1805. https://doi.org/10.1016/j.socscimed.2010.02.010

Clifton K (2003) Independent mobility among teenagers: exploration of travel to after-school activities. Transp Res Rec 1854:74–80. https://doi.org/10.3141/1854-08

Copperman RBA (2008) A comprehensive assessment of children's activity-travel patterns with implications for activity-based travel demand modeling. Dissertation, The University of Texas at Austin

Datz A, Cain A, Hamer P, Sibley-Perone J (2005) Teenage attitudes and perceptions regarding transit use. Technical report, National Center for Transit Research (NCTR), Florida

De Dios Ortuzar J, Willumsen LG (2001) Modelling transport. Wiley, New Jersey

Kamargianni M (2010) Exploring teenagers driving behavior in rural areas. In: Transport Research Arena 2010 conference, Brussels, Belgium

Kamargianni M (2014) Development of hybrid models of teenagers' travel behavior to school and to after-school activities. Dissertation, University of the Aegean

Line T, Chatterjee K, Lyons G (2010) The travel behaviour intentions of young people in the context of climate change. J Transp Geogr 18:238–246. https://doi.org/10.1016/j.jtrangeo.2009.05.001

Mackett RL (2013) Children's travel behaviour and its health implications. Transp Policy 26:66–72. https://doi.org/10.1016/j.tranpol.2012.01.002

Mackett R, Brown B, Gong Y, Kitazawa K, Paskins J (2007) Children's independent movement in the local environment. Built Environ 33:454–468. https://doi.org/10.2148/benv.33.4.454

Marzuki M, Mohamad J (2006) Mod pengangkutan ke sekolah: Satu kajian pilihan pelajar-pelajar sekolah menengah di Hulu Langat. http://eprints.uthm.edu.my/2729/1/17MOD_PENGANGKUTAN_KE_SEKOLAH_SATU_KAJIAN_PILIHAN_PELAJAR-P.pdf. Accessed 5 Mac 2018

McDonald NC, Deakin E, Aalborg AE (2010) Influence of the social environment on children's school travel. Prev Med 50:S65–S68. https://doi.org/10.1016/j.ypmed.2009.08.016

Nasrudin N, Nor ARM (2013) Travelling to school: transportation selection by parents and awareness towards sustainable transportation. Procedia Environ Sci 17:392–400. https://doi.org/10.1016/j.proenv.2013.02.052

Nishihara N, Belgiawan PF, Zürich IVTETH, Kim J, Schmöcker JD, Axhausen KW (2017) Identifying the relationship between parents' and child's car attitudes: for long-term management of car ownership. In: 17th Swiss transport research conference (STRC 2017), Monte Verita, Ascona. https://doi.org/10.3929/ethz-b-000130803

Paleti R, Copperman RB, Bhat CR (2011) An empirical analysis of children's after school out-of-home activity-location engagement patterns and time allocation. Transportation 38:273–303. https://doi.org/10.1007/s11116-010-9300-2

Storey P, Brannen J (2000) Young people and transport in rural areas. Joseph Rowntree Foundation. Youth Work Press, Leicester

Turner J, Pilling A (1999) Integrating young people into integrated transport: a community based approach to increase travel awareness. In: PTRC Young people and transport conference, Manchester

Chapter 3
Travel Behaviour Modification (TBM) Programmes

Abstract This chapter focuses on the travel behaviour modification (TBM) programmes as one of the recommended strategies to enhance the travel demand management methods. However, the quality of public transports has received higher priority instead, which explains why the TBM programmes are not widespread in the developing countries despite its influence in transforming the public behaviour towards sustainable transports. The examples of TBM programmes that have been done in developed countries are also summarized in this chapter. The identification towards the particular effects and outcomes from the programmes are able to provide insights for the development and implementation of TBM in the developing countries.

Keywords Travel behaviour modification · Intervention · Incentive Transportation planning · Travel demand management

3.1 What Is Travel Behaviour Modification (TBM)?

Essentially, TBM is a designed persuasive technique to realize a desired behaviour through incentives and persuasion based on the necessary information and specific goals or distinctive behavioural plan (Fujii and Kitamura 2003; Rose and Ampt 2003; Stopher and Bullock 2003; Taniguchi et al. 2003; Chatterjee and Bonsall 2009; Bamberg et al. 2011). With the promotion of public beliefs and knowledge, effective TBM in terms of communication presents positive prospects in realizing the desired behaviour and enhancing the public awareness and normative concern towards sustainable transports. Accordingly, the public may eventually attempt to travel with sustainable transports when they are provided with adequate information, personal assistance and motivation in overcoming the possible challenges (Cooper 2007). The public awareness and understanding of sustainable transports can be improved through information, which eventually initiate the attitudinal transformation in the travel behaviour (Steg and Vlek 2009). For instance, Wen et al. (2005) demonstrated the effectiveness of information delivery through

© The Author(s) 2019
N. S. A. Sukor and N. K. Basri, *Travel Behaviour Modification (TBM) Program for Adolescents in Penang Island*, SpringerBriefs on Case Studies of Sustainable Development, https://doi.org/10.1007/978-981-13-2505-2_3

banners, e-mail and flyers, which effectively reduced the total car drivers on road by 20% per week and lesser weekend trips within the inner city of Sydney, Australia.

On the other hand, the incentive-based technique can be solely applied in the attempt of encouraging the public to make use of public transports. This particular technique can also be combined with other techniques to increase the willingness to travel via public transports. For examples, Bartram (2009) recommended incentives in the form of financial assistance, such as cash payments, coupons, free public transport tickets or products. Thøgersen and Møller (2008) revealed increasing interest from a significant number of car drivers in using the public transports when they were provided with 1-month free pass and the schedule of public transports. A similar strategy was applied among the students in Kyoto University, which demonstrated the persistently high frequency of bus usage even after the valid period of the 1-month free pass (Fujii and Kitamura 2003). Thus, incentives can be an effective technique under the TBM programmes to encourage the use of public transports and reduce the dependency on private vehicles.

Nonetheless, Sheeran et al. (2005) propounded that firm motivation is necessary to prompt new behaviour even with a well-defined goal. In this case, this highlights the importance of behavioural intention in associating the established goal (of intention) with the commitment (Gollwitzer 1993) towards performing a specific behaviour (Fujii and Taniguchi 2005). The formation of behavioural intention reflects the combination of opportunities and strategic measures towards a specific goal in inducing specific behavioural change (Gärling and Rise 2002). Eriksson et al. (2008b) supported that planning a behavioural change or deliberately evaluating one's own travel behaviour would interrupt habitual travel mode choices. The implementation of intention and behavioural intention will be effective prior to behavioural planning (Gärling and Fujii 2002). Thus, it is pivotal to set a specific time, place and strategy for an effective implementation of TBM in prompting the required behavioural changes (Bamberg 2000; Fujii and Taniguchi 2005, 2006; Eriksson et al. 2008a). Moreover, it is imperative to grasp the effectiveness of specific TBM programmes in inducing the behaviour of using the public transports (Abrahamse and Matthies 2012; Sottile et al. 2015).

3.2 Examples of TBM Programmes

Table 3.1 presents the previous TBM programmes to provide insights into the development and implementation of future TBM programmes.

Table 3.1 Examples of communication-based interventions for TBM programmes

Author	Location	Target group	Intervention	Objective	Initiator	Data	Main findings
Individualized Marketing							
Brög et al. (2002)	South Perth (Australia)	15,000 households (35,000 people)	Direct dialogue (Mail or telephone)	To promote the use of public transports and non-motorized transports (cycling and walking)	Consultant company —'Socialdata'	Questionnaire (as the baseline data in 1997 and in the subsequent four months in 2000 when the programme ended)	Reduced car usage by 14% Increased carpool by 9%, walking by 35%, cycling by 61%, and public transports by 17% Walking captured half of the alternative modes for car trips
Travel Smart							
Rose and Marfurt (2007)	Victoria (Australia)	5577 participants	Ride to Work Day (RTWD) event in 2004 (offer comprehensive information on the available bicycle facilities and motivation via radio)	To promote cycling as an alternative travel option to the workplace	Victorian Department of Infrastructure (DOI)	Questionnaire (Data were compared between 2004 and 2005)	27% of the participants (first-timers) still opted for cycling five months after the end of the event Over 80% of these first-timers demonstrated willingness to cycle to their workplace; with 57% had influenced their decision to ride

(continued)

Table 3.1 (continued)

Author	Location	Target group	Intervention	Objective	Initiator	Data	Main findings
							The event significantly influenced the female participants compared to the male participants
Travel Blending							
Taylor and Ampt (2003)	Adelaide (Australia)	900 households	Living Neighbourhood® project	To reduce car usage by blending modes, blending activities and making sustainable changes on the regular basis	Monitoring process from the organisation and transport consultant company	Travel diaries in 1999	Reduced car usage by 14%, travel distance by 11%, and travel duration via car by 19%
Travel Feedback Programmes (TFP)							
Taniguchi and Fujii (2007a)	Sapporo (Japan)	398 participants (family members)	Participants were divided into group advice and group planning	To test the integrated process model	Researcher	3-day activity travel diary	The formation of behavioural intention and its translation into the actual behaviour were impeded by habit. Behaviour change was induced by

(continued)

Table 3.1 (continued)

Author	Location	Target group	Intervention	Objective	Initiator	Data	Main findings
							implementing intentions, which were influenced by behaviour intention The altruistic determinants influenced the behavioural intention
Fujii and Taniguchi (2005)	Sapporo (Japan)	292 students (of age 10 and 11 years)	155 students received individualized information, while 137 students were required to make behavioural plans	To reduce car usage	Researcher	Questionnaire (collected every three days—Sunday, Monday, and Tuesday)	Control group does not make significant changes Planning group: Reduced the trip duration by 27.7% and the car usage by 11.6% It is more effective to directly induce their behavioural intentions compared to offering incentives
Taniguchi and Fujii (2007b)	Obihiro (Hokkaido)	20,000 participants	Implemented mobility management (MM) measures	To promote an experimental community bus service (known as	Researcher	Questionnaire (conducted in the service areas and	Increased bus usage for the target group and no changes for the control group (indicated the

(continued)

Table 3.1 (continued)

Author	Location	Target group	Intervention	Objective	Initiator	Data	Main findings
			through the newsletter to promote bus service and free bus tickets	the 'Ring–Ring Bus')		monthly newsletters)	effectiveness of free bus tickets in promoting continuous bus usage) Increased bus usage by 34%: reading the newsletter strengthened the behavioural intention of using the bus in the future The word-of-mouth advertising prompted the participants to provide recommendations on using the bus, which encouraged bus usage among those who received these recommendations
Tørnblad et al. (2014)	Norway	2000 employees of 6 companies	To introduce treatments —'tailored information' and 'free transit pass'	To test whether tailored information on local public transport options and free pass encourage the	Researcher	1st questionnaire (on mode choice) 2nd questionnaire and 3rd questionnaire (the decision to switch the housing	Both treatments did not reduce the usage of private cars to the workplace in the short- and medium-term analysis

(continued)

Table 3.1 (continued)

Author	Location	Target group	Intervention	Objective	Initiator	Data	Main findings
				commuters to opt for public transport		location or workplace and the usage of the free transit pass)	Despite the high taxes and fuel price, the private cars were preferred due to the inconvenience of public transports and non-motorized transports (cycling and walking) during the winter season. Limitation: Require controlled experiments on the effectiveness of MM measures

Personalized Travel Plans

Author	Location	Target group	Intervention	Objective	Initiator	Data	Main findings
van Bladel et al. (2009)	Flanders (Belgium)	2500 households with 272 respondents	Activity scheduling and rescheduling	To analyse the factors that influence the actual activity scheduling process using detailed activity travel data from an extensive data set	Researcher	7-day activity travel schedule	The activity and schedule characteristics significantly influenced the activity planning. The rescheduling model also has highly significant activities and schedule attributes. Despite the statistical evidence that demonstrated a

(continued)

Table 3.1 (continued)

Author	Location	Target group	Intervention	Objective	Initiator	Data	Main findings
							significant influence of individual-specific preferences for planning and rescheduling towards the process of activity scheduling, most individual and household attributes considered as did not influence the activity planning and rescheduling behaviour
Meloni et al. (2013)	Cagliari (Italy)	85 participants	The implementation of a personalized travel planning (PTP) to promote the use of light metro	To analyse the propensity of individuals for travel change and to model the underlying factors that contribute to behavioural change	Researcher	Two-week panel survey personalized travel plan after the first week of observation (before) and the second week to monitor the post-behaviour (after)	Increase in the Park and Ride (P&R) by 10% and car users by 10%. Car users demonstrated a higher willingness to pay for shorter travel time compared to P&R users. Those who travelled more than 25,000 km annually demonstrated lower likelihood to

(continued)

Table 3.1 (continued)

Author	Location	Target group	Intervention	Objective	Initiator	Data	Main findings
							change their preferred transport option Those with more work trips were more likely to change travel behaviour Individual habit exhibited higher probability to change travel behaviour

Voluntary Change Strategies

| Taniguchi et al. (2014) | Värmland (Sweden) | 321 car commuters | The implementation of free monthly travel card for public transports (valid for a certain period only) | To promote the use of public transports between home and workplace at least three times per week for four weeks | Researcher | Questionnaire (satisfaction with travel scale (STS): to measure the travel experience based on the rating of the goal achievement) | The distance between home and the nearest bus stop contributed significantly negative effects on STS
The STS contributed significantly positive effects on goal achievement
Both goal achievement and STS contributed significantly positive effects on the use of |

(continued)

Table 3.1 (continued)

Author	Location	Target group	Intervention	Objective	Initiator	Data	Main findings
							public transports and future goals Both goal achievement and travel experience were important for voluntary behavioural change
Thøgersen (2014)	Greater Copenhagen	1071 randomly sampled car users	Car users received free monthly travel card for public transports with a customized travel plan or planning intervention (a control group received no intervention)	To promote the use of public transports in a field experiment based on a solid behavioural–theoretical framework	Researcher	Questionnaire (attitudinal variables, car habit, and travel behaviour were measured before, after the intervention and in the following six months)	Free monthly travel card significantly increased the use of public transports The effect was mediated through a change in behavioural intention, rather than a change in perceived constraint Although the effect was weaker following the expiry of free monthly travel card for public transports, the effect remained evident five months later

(continued)

Table 3.1 (continued)

Author	Location	Target group	Intervention	Objective	Initiator	Data	Main findings
Commitment Improvement Measures							
Abou-Zeid et al. (2012)	Switzerland	30 participants	Free monthly travel card for public transports without information on the public transportation schedule	To assess the dynamics of the travel satisfaction ratings before and after the intervention	Researcher	Questionnaire (pre-treatment, treatment, and post-treatment weeks) The participants were required to commute using the public transports for at least 2–3 days per week	None of the participants demonstrated complete in their travel behaviour with only a few of them occasionally opted for public transports Participants revealed significantly higher satisfaction level in using cars during the post-treatment week Participants revealed higher satisfaction level in using public transports, which exceeded their expectation

References

Abou-Zeid M, Witter R, Bierlaire M, Kaufmann V, Ben-Akiva M (2012) Happiness and travel mode switching: findings from a Swiss public transportation experiment. Transp Policy 19:93–104. https://doi.org/10.1016/j.tranpol.2011.09.009

Abrahamse W, Matthies E (2012) Informational strategies to promote pro-environmental behaviour: changing knowledge, awareness and attitudes. Environ Psychol 223–232. https://www.researchgate.net/profile/Ellen_Matthies/publication/285668068_Informational_strategies_to_promote_pro-environmental_behaviours_Changing_knowledge_awareness_and_attitudes/links/56c1bb5808ae44da37fea2aa.pdf. Accessed 15 May 2018

Bamberg S (2000) The promotion of new behavior by forming an implementation intention: results of a field experiment in the domain of travel mode choice. J Appl Soc Psychol 30:1903–1922. https://doi.org/10.1111/j.1559-1816.2000.tb02474.x

Bamberg S, Fujii S, Friman M, Gärling T (2011) Behaviour theory and soft transport policy measures. Transp Policy 18:228–235. https://doi.org/10.1016/j.tranpol.2010.08.006

Bartram A (2009) Behaviour change intervention tools. Discussion Paper, Government of South Australia. http://www.dpti.sa.gov.au/__data/assets/pdf_file/0003/42627/Review_of_Behaviour_Change_Intervention_Tools.pdf. Accessed 30 Jun 2018

Brög W, Erl E, Mense N (2002) Individualised marketing changing travel behaviour for a better environment. In: OECD workshop: environmentally sustainable transport, vol 5, pp 06–12. http://www.socialdata.de/info/IndiMark.pdf. Accessed 3 Aug 2018

Chatterjee K, Bonsall P (2009) Editorial for special issue on evaluation of programmes promoting voluntary change in travel behaviour. Transp Policy 16:279–280. https://doi.org/10.1016/j.tranpol.2009.10.001

Cooper C (2007) Successfully changing individual travel behavior: Applying community-based social marketing to travel choice. Transp Res Rec 2021:89–99. https://doi.org/10.3141/2021-11

Eriksson L, Friman M, Gärling T (2008a) Stated reasons for reducing work-commute by car. Transp Res Part F Traffic Psychol Behav 11:427–433. https://doi.org/10.1016/j.trf.2008.04.001

Eriksson L, Garvill J, Nordlund AM (2008b) Interrupting habitual car use: the importance of car habit strength and moral motivation for personal car use reduction. Transp Res Part F Traffic Psychol Behav 11:10–23. https://doi.org/10.1016/j.trf.2007.05.004

Fujii S, Kitamura R (2003) What does a one-month free bus ticket do to habitual drivers? An experimental analysis of habit and attitude change. Transportation 30:81–95. https://doi.org/10.1023/A:1021234607980

Fujii S, Taniguchi A (2005) Reducing family car-use by providing travel advice or requesting behavioral plans: an experimental analysis of travel feedback programs. Transp Res Part D Transp Environ 10:385–393. https://doi.org/10.1016/j.trd.2005.04.010

Fujii S, Taniguchi A (2006) Determinants of the effectiveness of travel feedback programs—a review of communicative mobility management measures for changing travel behaviour in Japan. Transp Policy 13:339–348. https://doi.org/10.1016/j.tranpol.2005.12.007

Gärling T, Fujii S (2002) Structural equation modeling of determinants of planning. Scand J Psychol 43:1–8. https://doi.org/10.1111/1467-9450.00263

Gärling T, Rise J (2002) Understanding attitude, intention, and behavior: a common interest to economics and psychology. In: Spash CL, Biel A (eds) Social psychology and economics in environmental research. Cambridge University Press, Cambridge

Gollwitzer PM (1993) Goal achievement: the role of intentions. Eur Rev Soc Psychol 4:141–185. https://doi.org/10.1080/14792779343000059

Meloni I, Sanjust B, Sottile E, Cherchi E (2013) Propensity for voluntary travel behavior changes: an experimental analysis. Procedia Soc Behav Sci 87:31–43. https://doi.org/10.1016/j.sbspro.2013.10.592

Rose G, Ampt E (2003) Travel behavior change through individual engagement. In: Handbook of transport and the environment. Emerald Group Publishing Limited

Rose G, Marfurt H (2007) Travel behaviour change impacts of a major ride to work day event. Transp Res Part A Policy Pract 41:351–364. https://doi.org/10.1016/j.tra.2006.10.001

Sheeran P, Webb TL, Gollwitzer PM (2005) The interplay between goal intentions and implementation intentions. Pers Soc Psychol B 31:87–98. https://doi.org/10.1177/0146167204271308

Sottile E, Cherchi E, Meloni I (2015) Measuring soft measures within a stated preference survey: the effect of pollution and traffic stress on mode choice. Transp Res Procedia 11:434–451. https://doi.org/10.1016/j.trpro.2015.12.036

Steg L, Vlek C (2009) Encouraging pro-environmental behaviour: an integrative review and research agenda. J Environ Psychol 29:309–317. https://doi.org/10.1016/j.jenvp.2008.10.004

Stopher PR, Bullock P (2003) Travel behaviour modification: a critical appraisal. In: Australasian Transport Research Forum (ATRF), Wellington, New Zealand

Taniguchi A, Fujii S (2007a) Process model of voluntary travel behavior modification and effects of travel feedback programs. Transp Res Rec J Transp Res B 2010:45–52. https://doi.org/10.3141/2010-06

Taniguchi A, Fujii S (2007b) Promoting public transport using marketing techniques in mobility management and verifying their quantitative effects. Transportation 34:37–49. https://doi.org/10.1007/s11116-006-0003-7

Taniguchi A, Hara F, Takano SE, Kagaya SI, Fujii S (2003) Psychological and behavioral effects of travel feedback program for travel behavior modification. Transp Res Rec 1839:182–190. https://doi.org/10.3141/1839-21

Taniguchi A, Grääs C, Friman M (2014) Satisfaction with travel, goal achievement, and voluntary behavioral change. Transp Res Part F Traffic Psychol Behav 26:10–17. https://doi.org/10.1016/j.trf.2014.06.004

Taylor MA, Ampt ES (2003) Travelling smarter down under: policies for voluntary travel behaviour change in Australia. Transp Policy 10:165–177. https://doi.org/10.1016/S0967-070X(03)00018-0

Thøgersen J (2014) Social marketing in travel demand management. Handbook of sustainable travel. Springer, Netherlands, pp 113–129

Thøgersen J, Møller B (2008) Breaking car use habits: the effectiveness of a free one-month travelcard. Transportation 35:329–345. https://doi.org/10.1007/s11116-008-9160-1

Tørnblad SH, Kallbekken S, Korneliussen K, Mideksa TK (2014) Using mobility management to reduce private car use: results from a natural field experiment in Norway. Transp Policy 32:9–15. https://doi.org/10.1016/j.tranpol.2013.12.005

van Bladel K, Bellemans T, Janssens D, Wets G (2009) Activity travel planning and rescheduling behavior: empirical analysis of influencing factors. Transp Res Rec: J Transp Res Rec 2134:135–142. https://doi.org/10.3141/2134-16

Wen LM, Orr N, Bindon J, Rissel C (2005) Promoting active transport in a workplace setting: evaluation of a pilot study in Australia. Health Promot Int 20:123–133. https://doi.org/10.1093/heapro/dah602

Chapter 4
Intervention Programmes in Penang Island

Abstract The current chapter proposes several intervention strategies to promote sustainable transports among the adolescents (secondary school students) in Penang Island. Overall, the introduced travel behaviour modification (TBM) programme, which focused on the adolescents, aimed to promote awareness on the implications of carbon emissions from vehicles as well as to encourage the use of sustainable transports. In this study, the interventions were included motivation, journey planning and incentive that focused on four groups of adolescents. The intervention programmes were conducted through a campaign where the adolescents were provided with campaign kits to encourage them to commit with the programmes. The proposed interventions are expected to contribute to the planning and implementation of any behavioural modification programmes.

Keywords Travel behaviour programme · Adolescents · Carbon footprint Motivation · Incentive · Travel diary

4.1 Let's Reduce Carbon Footprint Campaign

With the collaboration between the School of Civil Engineering and the primary bus operator in Penang Island (Rapid Penang), 'Let's Reduce Carbon Footprint Campaign' was launched in June 2014 for 3 months as part of the initiation of the TBM programme. More information on Rapid Penang are accessible online (Link: http://www.rapidpg.com.my/). The campaign aimed to educate these adolescents on the environmental implications of carbon emissions. Through this campaign, the plausible strategies and directions were assessed for the extensive use of private vehicles, which has contributed to the prevalent traffic congestion issue in Penang Island. Prior to the launch of the campaign, the selection of the targeted participants, the types of intervention, and the preparation of campaign materials were established. The following sub-chapters discuss these specific considerations in detail.

© The Author(s) 2019
N. S. A. Sukor and N. K. Basri, *Travel Behaviour Modification (TBM) Program for Adolescents in Penang Island*, SpringerBriefs on Case Studies of Sustainable Development, https://doi.org/10.1007/978-981-13-2505-2_4

4.2   Selection of Participants

Essentially, an appropriate targeted audience must be ascertained for the success of a programme. Chapter 2 rationalized the need to target the adolescents for the campaign given their daily travel need to a fixed destination (school) on the daily basis throughout the week. Thus, the adolescents were meticulously selected as the targeted participants for the programme through their school based on a series of specific procedure. The procedure was necessary in order to attain the formal approval from the school and to prevent any conflicting schedule between the school activities of these adolescents and the implemented programme.

First, the targeted public secondary schools for this programme were mainly selected based on their accessibility to the bus service: (1) the school is within the scheduled bus route and (2) a bus stop is located within 500 m from the main entrance of the school. The information on the accessibility to the bus service, especially on the bus routes, were corroborated with Rapid Penang. Overall, the bus service in Penang Island reflects good accessibility, but it still needs extensive improvement to solve the first and last mile accessibility issues. Meanwhile, the distance of 500 m was considered based on the evidence provided by Azmi and Karim (2012), which revealed that Malaysians were only willing to walk within 200 m between the urban neighbourhood and community facility. Therefore, these specific criteria narrowed the selection of schools for the programme. The accessibility to the bus service was considered as an important attribute in the selection of schools to accurately assess the effects of TBM programme. It is imperative to establish a supportive setting for the adolescents to conveniently use the bus service. As a result, only 8 schools from a total of 48 public secondary schools satisfied the required criteria, which are revealed in Fig. 4.1.

Second, the formal approval from the Ministry of Education (MOE), the Penang Education Department, and the principal of the selected schools were acquired. This was to ensure that the introduced programme did not put the adolescents at risk in any physical or psychological forms. Following the approval, several teachers in every school were assigned to be part of the programme. Fundamentally, the TBM programme supports voluntary behavioural changes among the participants. The students in the selected schools were informed about the nature of the programme and subsequently invited to voluntarily participate.

A total of 200 students initially demonstrated interest to participate, but only 176 students agreed to commit to the proposed interventions of the programme. Specifically, there were 54 male participants and 122 female participants, which indicated that the female adolescents have a higher propensity for voluntary participation. However, certain studies demonstrated that the gender has insignificant effects towards school travel behaviour (Valentine and McKendrck 1997; Irawan and Sumi 2011; Limanond et al. 2011; McDonald 2012). Regardless, the small sample size did not pose any issue; the number of participants is inevitably lower since the behavioural experiments require longer time. After all, the minimum recommended sample size for pre–post experiment is between 20 and 30

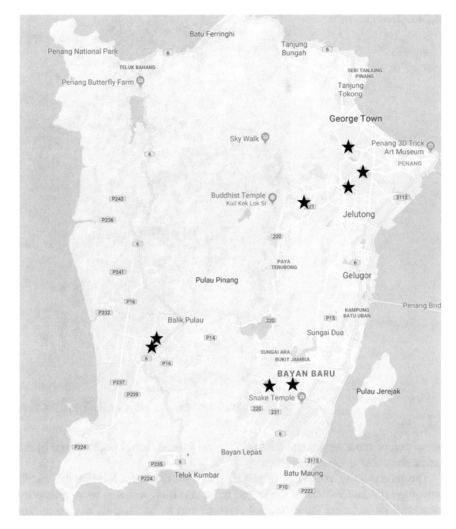

Fig. 4.1 Location of selected schools for TBM programme

participants per condition (Cozby 2006). Accordingly, Fujii and Kitamura (2003), Taniguchi et al. (2003), and Fujii and Taniguchi (2005, 2006) also employed a similarly small sample size (<100).

4.3   Campaign Materials

The campaign materials aimed to motivate the adolescents to commit to the programme. The participating students, as the campaign ambassadors, were provided with 'Travel Kit Folder' (Fig. 4.2), in the first assembly. Each campaign kit

Fig. 4.2 Travel Kit Folder

comprised of the motivational pamphlet, travel journal, travel pocket map, button badge and other materials.

The motivational pamphlet (Fig. 4.3) provided the basic information on the environmental implications of the carbon emissions from vehicles, while promoting awareness on reducing the individual carbon footprint in the daily commute. The two-page pamphlet briefly described travel-related carbon footprint, a basic guideline on reducing individual carbon footprint and general information on global warming effects.

Meanwhile, the participants were provided with the pictorial bus routes, the bus fare structure and the specific operating bus numbers (for various places of interest) in Penang Island through the provided travel pocket map (Fig. 4.4), which was sponsored by Rapid Penang (Prasarana Malaysia Berhad). It should be noted that any similar programme should seek for the new version of the travel pocket map since there have been changes to the existing routes.

Besides button badge, the campaign kit also included plastic hand fan. These motivational gifts were specially designed with the campaign message of 'Jom ride the bus, cycling, and walking for a better tomorrow' and the pictorial images of bus,

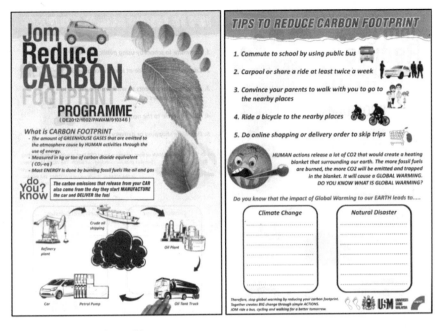

Fig. 4.3 Motivational pamphlet

Fig. 4.4 Travel pocket map

Fig. 4.5 Button badge

cycling, and walking. The term 'jom' denotes 'let us together'. As the campaign ambassadors, the participants were encouraged to display the button badge on their school bags at all times, especially during their bus ride (Fig. 4.5).

Last but not least, the travel journal, which was one of the most significant materials for the TPB programme, comprised of three sections. In particular, the participants were required to provide their school identification, school address, home address and information on the main transportation mode of household (i.e. type of vehicle, vehicle engine displacement (cc), choice of fuel type, fuel consumption (litre per 100 km), year of production) in Section A.

In general, Section B provides information on the participants' daily travel pattern and the carbon emissions from the main transportation mode of the household. Specifically, the participants were required to report their daily travel, such as the purpose of travel, the origin and destination of the travel, the departure and arrival time, the distance travelled and the mode of transport. Meanwhile, Section C (provided with additional sheets after the final page of the travel journal) was specifically designed for the participants who were required to plan their daily travel (journey planning intervention). Thus, not all participants were required to complete this particular section. Figure 4.6 presents an example of information to be provided in the travel journal.

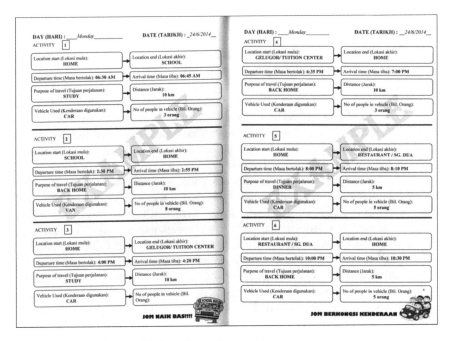

Fig. 4.6 Example of required information in the travel journal

4.4 Intervention Programmes

The participants were first briefed on the concept of carbon footprint and how the carbon footprint is calculated for them to fully embrace the assigned tasks of intervention programmes. During this specially organized session, several online sources were recommended for the participants to explore. For example, the website under the Tree for Life that provides a basic carbon footprint calculator (Link: https://www.treesforlife.org.au/kids-carbon-calculator). The obtained result provides the amount of carbon (in kilograms) produced from their daily or weekly activities.

The participants were also prompted to explore advanced information on carbon footprint and pointers on how to reduce the individual carbon footprint (e.g. Link: https://www.carbonfootprint.com/calculator.aspx). Basically, the calculation of carbon footprint requires values (in USD) of monthly electrical bill in a household, total expenditures (such as education, food, insurance, medical or others) and monthly transport usage, which revealed the total offset of carbon footprint per individual (of CO2e unit in tonnes). The data were exclusively used for the briefing session only.

Besides that, the participants were guided on how they should provide the necessary details on their daily travel in the travel journal (denoted as TJ1) throughout the week (of both weekdays and weekends). After 1 week, the participants were reminded to submit their travel journal for evaluation. Following that, the selected schools were divided into four groups (two schools per group) for the intervention programmes, namely control group (G1), journey planning group (G2), incentive group (G3) and journey planning with incentive group (G4). The participants in G1 did not undergo any intervention, except for motivation. Meanwhile, the participants in G2 were required to plan their daily travel for the week on the provided additional sheets (denoted as TJ2), which were to be submitted for evaluation at the end of the week. One-month unlimited free bus passes for Rapid Penang were provided to the participants in G3. Lastly, interventions of both G2 and G3 were combined for G4.

All participants gathered for the feedback phase to obtain data on the individual carbon footprint at the end of the week. Specifically, the details of their daily travel were sorted and analysed. The participants were notified on their total carbon footprint for the week before and after the intervention programmes. The individual carbon footprints based on the information provided in TJ1 and TJ2 were compared and revealed to each participant. Participants who demonstrated positive changes in their travel behaviour based on the results of the individual carbon footprint were formally acknowledged and encouraged to sustain these positive changes. Figure 4.7 illustrates the overall flow of the intervention programmes for the campaign.

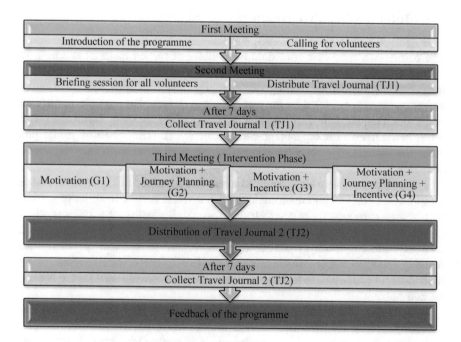

Fig. 4.7 Flow of intervention programmes

4.4.1 Motivation

The participants in all intervention groups received similar motivation session, which mainly served to promote awareness on the significance of low-carbon mobility and its positive contributions to the environment and society. The negative implications of carbon emissions from vehicles as well as how air pollution worsens the effects of global warming and climate change were described to the participants. In addition, they were also introduced to the various alternative travel options (as opposed to private vehicles) to realize low carbon mobility. Sustainable transports, such as the public transports (e.g. bus) and non-motorized transports (e.g. cycling and walking), were highlighted as the recommended travel options. Apart from that, the participants were required to discuss the effects of carbon emissions from the motor vehicles towards climate change and global warming based on the information provided on the distributed motivational pamphlet. The group discussion mainly aimed to transform the existing perspectives towards public transports among the participants. They were also prompted to adapt their travelling behaviour according to the provided guideline on reducing individual carbon footprint.

4.4.2 Journey Planning

After the motivation session, the participants in G2 and G4 were invited to another session. For this session, the participants were prompted to access an online planner, which was provided by Rapid Penang (Link: http://www.rapidpg.com.my/journey-planner/plan-my-trip). Referring to the online planner (known as 'Journey Planner'), the participants were guided on how to plan their 7-day travel using TJ2.

The participants in both groups were instructed to modify their travel behaviour and commit to their daily travel plan. In particular, they were expected to travel by bus throughout the week. The participants were also reminded to make use of their travel pocket map. Both Rapid Penang website and the distributed travel pocket map presented the details of bus routes (including the estimated departure and arrival time), the bus fare structure, the number of transits, the travel distance, the specific operating bus numbers for various destinations in Penang Island. Figure 4.8 displays the journey planning session.

4.4.3 Incentive

Meanwhile, the participants in G3 and G4 were individually provided with 1-month unlimited free bus pass (Fig. 4.9) after the motivation session. In particular, they were encouraged to ride the Rapid Penang bus for a month. With the opportunity to personally experience the bus ride for a month, the intervention was expected to

Fig. 4.8 The travel planning activities for the intervention groups—G2 and G4

(a) Front cover of the bus pass   (b) Back cover of the bus pass

Fig. 4.9 Example of 1-month unlimited free bus pass

draw a positive attitude among the participants towards public transports. Accordingly, the intervention is considered as a temporary structural change strategy, which is typically expressed as a marketing strategy, to promote a specific product for market expansion. Similarly, the participants in G3 and G4 were also reminded to make use of their travel pocket map in their daily travel using the Rapid Penang bus.

References

Azmi DI, Karim HA (2012) A comparative study of walking behaviour to community facilities in low-cost and medium cost housing. Procedia Soc Behav Sci 35:619–628. https://doi.org/10.1016/j.sbspro.2012.02.129

Cozby PC (2006) Method in behavioural research, 9th edn. McGraw-Hill International Edition

Fujii S, Kitamura R (2003) What does a one-month free bus ticket do to habitual drivers? An experimental analysis of habit and attitude change. Transportation 30:81–95. https://doi.org/10.1023/A:1021234607980

Fujii S, Taniguchi A (2005) Reducing family car-use by providing travel advice or requesting behavioral plans: an experimental analysis of travel feedback programs. Transp Res Part D Transp Environ 10:385–393. https://doi.org/10.1016/j.trd.2005.04.010

Fujii S, Taniguchi A (2006) Determinants of the effectiveness of travel feedback programs—a review of communicative mobility management measures for changing travel behaviour in Japan. Transp Policy 13:339–348. https://doi.org/10.1016/j.tranpol.2005.12.007

Irawan MZ, Sumi T (2011) Promoting active transport in students' travel behavior: a case from Yogyakarta (Indonesia). J Sustain Dev 4:45–52. https://doi.org/10.5539/jsd.v4n1p45

Limanond T, Butsingkorn T, Chermkhunthod C (2011) Travel behaviour of university students who live on campus: a case study of a rural university in asia. Transp Policy 18:163–171. https://doi.org/10.1016/j.tranpol.2010.07.006

McDonald NC (2012) Is there a gender gap in school travel? An examination of US children and adolescents. J Transp Geogr 20:80–86. https://doi.org/10.1016/j.jtrangeo.2011.07.005

Taniguchi A, Hara F, Takano SE, Kagaya SI, Fujii S (2003) Psychological and behavioral effects of travel feedback program for travel behavior modification. Transp Res Rec 1839:182–190. https://doi.org/10.3141/1839-21

Valentine G, McKendrck J (1997) Children's outdoor play: exploring parental concerns about children's safety and the changing nature of childhood. Geoforum 28:219–235. https://doi.org/10.1016/S0016-7185(97)00010-9

Chapter 5
Implications of Travel Behaviour Modification (TBM) Programme

Abstract This chapter explores the implications of the intervention programmes under the TBM programme among the adolescents in the aspects of the purpose of the travel, time allocation for daily activities, hourly trip frequencies, choice of transport for travel, travel distance, travel duration and carbon emissions from vehicles. It should be noted that the discussed findings on the implications of the intervention programmes were based on a short-term period only. The implementation of intervention programmes were found to significantly reduce the amount of carbon emission from vehicles during the weekdays. As for the weekend travel, the adolescents that received motivation and incentives reported significant reduction in the amount of carbon emission compare to the other intervention groups.

Keywords Trip frequency · Travel purpose · Mode choice · Travel time
Travel distance · Carbon emission

5.1 Purpose of Travel

The purpose of travel denotes the motive of the trip made. Table 5.1 classifies all purposes of travel among the participants. According to the submitted travel journals (denoted as TJ1) of 176 participants, there were 2646 trips made during the weekdays and 916 trips made during the weekends. Meanwhile, Fig. 5.1 displays the proportion of various purposes of travel on weekdays and weekends among the participants in the pre-intervention programmes. On average, the participants made 3.01 trips per day on weekdays and 2.59 trips per day on weekends. The trips made on weekdays were mainly to return home (47.9%), which were followed by trips to school (33.3%) and for extra academic class (6.3%). Apart from home, the participants mostly travelled to school for regular school activities or attended extra academic class at school or academic centre. These participants demonstrated higher propensity to make a trip back home between destinations. There were fewer trips made for leisure (4.7%), personal business (3.2%), dining out (2.3%), sports

© The Author(s) 2019
N. S. A. Sukor and N. K. Basri, *Travel Behaviour Modification (TBM) Program for Adolescents in Penang Island*, SpringerBriefs on Case Studies of Sustainable Development, https://doi.org/10.1007/978-981-13-2505-2_5

Table 5.1 Classification of the purpose of travel

Purpose of travel	Description
School (SC)	Trip to school
Extra academic class (EA)	Trip after formal school hours (either back to school or academic centre)
Visit relatives or friends (V)	Trip to relative's or friend's house
Personal business (P)	Trip to bank, clinic or hospital, religious place (mosque, church, or temple), or any place for running errands
Sports (SP)	Trip to any place for sports or other recreational activities
Eat (E)	Trip to café or restaurant for a meal (breakfast, lunch, teatime, dinner, supper or takeaway)
Leisure (L)	Trip to any social place for shopping, meeting up with friends or movie
Return home (H)	Trip to home (either as a brief stop location before the next trip after any listed purpose of travel or as the final destination for the day)

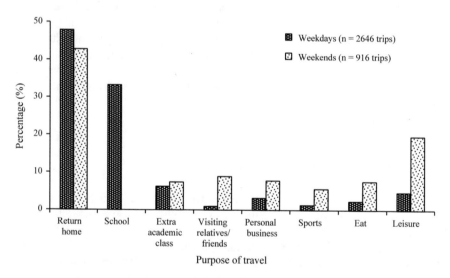

Fig. 5.1 Purposes of the trip in pre-intervention programmes

(1.3%) and visiting relatives or friends (1.0%) among the participants given their fixed time at school.

Similarly, most of the trips were to return home (42.8%) on weekends, as well. Majority of the participants made a trip back home between destinations, which corroborated that most trips exhibited simple trip chain. Apart from that, most of the trips made on weekends among the participants were for leisure (19.7%), which reaffirmed no school-related activities over the weekend. Nevertheless, there were lesser trips for visiting relatives or friends (9.0%), personal business (7.9%), dining out (7.6%), extra academic class (7.4%) and sports (5.7%). Overall, the results showed that the adolescents involved with more trips during weekdays compared to weekends.

Figure 5.2 presents the proportion of various purposes of travel on weekdays and weekends in the post-intervention programmes. Referring to the collected travel journal (TJ2), the participants made 2350 trips made on weekdays and 543 trips on weekends. On average, there were 2.67 trips per day on weekdays and 1.54 trips per day on weekends. Most of the trips made on weekdays were to return home (46.6%). The second highest trip made among the participants was travelling to school (37.4%) given their mandatory fixed time at school. Apart from extra academic class (6.2%) to improve their studies, there were fewer trips for leisure (2.5%), personal business (2.5%), dining out (1.4%), sports (1.1%), and visiting relatives or friends (0.5%). The results demonstrated the propensity of returning home after an activity or between activities among the participants, which reaffirmed the manifestation of simple trip chains or a sequence of trips with a brief stop at home, as one of the destinations between other destinations (for outdoor activities) in the daily travel.

The trip to return home (42.8%) was also the highest on weekends in the post-intervention programmes. Since there is no school over the weekend, the trip for leisure (14.5%) was evidently the next highest, which was followed by extra academic class (10.7%), personal business (9.4%), dining out (7.9%), visiting relatives or friends (5.9%). However, only a few participants enjoy sports, which contributed to the lowest percentage of trip made for sports (5.0%) over the weekend.

Subsequently, the average trip frequency on weekdays and weekends in the post-intervention programmes were compared to that of the pre-intervention programmes according to the four intervention groups. The results are illustrated in Fig. 5.3.

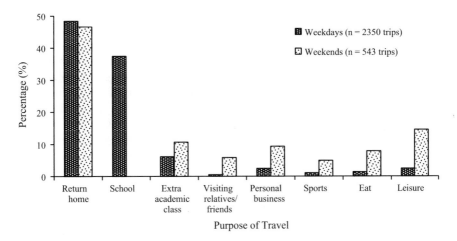

Fig. 5.2 Purposes of the trip in post-intervention programmes

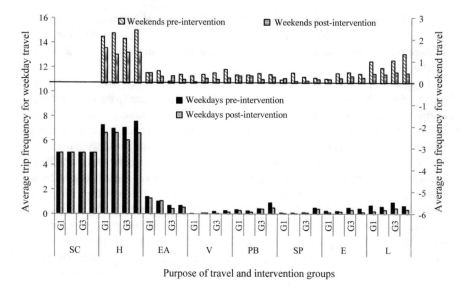

Fig. 5.3 Average trip frequency for pre-intervention and post-intervention programmes during weekdays and weekends

First, there were similar number of trips made to school on weekdays among the participants in the pre-intervention and post-intervention programmes. After all, these participants are obligated to attend school on weekdays. Second, the average trip frequency for returning home was reduced following the implementation of intervention programmes. Third, there were no significant differences for the trips made for extra academic class and sports among the participants despite the implementation of intervention programmes. Apart from the compulsory school attendance, the most common activities among the participants were attending these extra academic classes or playing sports.

Following that, the paired-sample t-test analysis was performed for the obtained data between the pre-intervention and post-intervention programmes, which revealed significant differences (Table 5.2) for G1 ($t = 2.661$, df = 43, $p < 0.05$), G3 ($t = 4.517$, df = 43, $p < 0.001$), and G4 ($t = 3.960$, df = 43, $p < 0.001$) when it comes to home as the destination for the weekday travel. Accordingly, G3 recorded the **highest mean difference** ($M_{diff} = 1.023$, SD = 1.501), followed by G4 ($M_{diff} = 0.955$, SD = 1.599), and G1 ($M_{diff} = 0.636$, SD = 1.586).

As for G2, the number of trips demonstrated a slight decrease following the implementation of intervention programmes; thus, resulting to statistically insignificant mean difference. Similar results were also revealed for leisure trips—G1 ($t = 2.575$, df = 43, $p < 0.05$), G3 ($t = 2.908$, df = 43, $p < 0.05$) and G4 ($t = 2.108$, df = 43, $p < 0.05$) revealed statistically significant differences. Specifically, both G1 (SD = 1.223) and G3 (SD = 1.089) recorded high mean difference of 0.477, while G4 ($M_{diff} = 0.955$, SD = 1.599) exhibited the smallest significant mean difference. However, when it comes to visiting relatives or friends,

Table 5.2 Paired-sample *t*-test results of trip frequency in the purpose of travel according to intervention groups

Purposes of travel	G1		G2		G3		G4	
	M_{diff}	*t*-value	M_{diff}	*t*-value	M_{diff}	*t*-value	M_{diff}	*t*-value
Weekday travel								
H	**0.636**	**2.661**	0.341	1.211	**1.023**	**4.517**	**0.955**	**3.960**
EA	0.114	0.616	−0.045	−0.313	0.250	1.806	0.136	0.829
V	0.045	1.431	0.000	0.000	**0.159**	**2.201**	0.114	1.301
PB	0.068	0.534	0.091	1.431	0.023	0.144	**0.409**	**2.663**
SP	0.045	1.431	0.068	1.138	0.023	0.330	0.091	0.599
E	0.136	1.634	0.045	0.388	0.182	1.386	**0.295**	**2.232**
L	**0.477**	**2.575**	0.273	1.906	**0.477**	**2.908**	**0.295**	**2.108**
Weekend travel								
H	**0.523**	**2.821**	**0.977**	**5.632**	**0.636**	**2.712**	**1.023**	**5.793**
EA	0.000	0.000	**0.250**	**2.545**	−0.250	−2.881	**0.227**	**2.226**
V	**0.273**	**2.901**	0.182	1.666	0.295	1.643	**0.386**	**2.643**
PB	0.045	0.496	0.045	0.340	**0.273**	**2.606**	0.114	1.000
SP	−0.068	−1.354	**0.386**	**3.269**	**0.182**	**2.074**	0.068	0.903
E	0.023	0.374	0.227	1.462	0.182	1.745	0.182	1.596
L	**0.568**	**4.316**	**0.295**	**2.003**	**0.545**	**3.464**	**0.886**	**4.960**

Notes Highlighted data in bold denotes statistical significance at 0.05 level; *G1* represents the control group; *G2* represents the journal planning group; *G3* represents the incentive group; *G4* represents the journal planning with incentive group; *H* denotes home; *EA* denotes extra academic class; *V* denotes visit relatives or friends; *PB* denotes personal business; *SP* denotes sports; *E* denotes eat; *L* denotes leisure

only G3 ($M_{\text{diff}} = 0.159$, SD = 0.479) demonstrated statistically significant difference ($t = 2.201$, df = 43, at $p < 0.05$). Meanwhile, G4 demonstrated statistically significant difference for trips for personal business ($t = 2.663$, df = 43, $p < 0.05$) and dining out ($t = 2.232$, df = 43, $p < 0.05$) in the attempt to reduce carbon footprint.

Overall, these participants reduced their trips for their weekday travel following the implementation of intervention programmes based on the average trip frequency data despite the results of statistically insignificant mean differences for certain groups. These results supported the prior findings on the trip chaining pattern of staying at home after regular school hours and the increase in the number of simple trip chains following the implementation of intervention programmes among the participants (G1, G3 and G4). Additionally, the participants demonstrated the tendency to minimize the number of unnecessary trips on weekdays in their attempt to reduce their carbon footprint.

On the other hand, the participants also demonstrated efforts of reducing the number of weekend trips for most of the travel purposes, with the exception of trips for the purposes of extra academic class and sports. The obtained results demonstrated mixed findings in the overall trips for extra academic class between pre-intervention and post-intervention programmes across groups. First, there were significant differences between the pre-intervention and post-intervention programmes for all groups, except G1. Second, the number of trips significantly decreased in G2 ($t = 2.545$, df = 43, $p < 0.05$) and G4 ($t = 2.226$, df = 43, $p < 0.05$). Third, the number of trips significantly increased in G3 ($t = -2.881$, df = 43, $p < 0.05$). When it comes to sports, only participants in G2 ($t = 3.2649$, df = 43, $p < 0.05$) and G3 ($t = 2.074$, df = 43, $p < 0.05$) statistically significantly reduced the overall trips over the weekend.

Apart from that, the overall trips made for dining out among the participants in the pre-intervention and post-intervention programmes showed no significant differences in mean. After all, the participants were likely to have meals with their family at home over the weekend. Meanwhile, the numbers of trips to return home and for leisure activities across groups were statistically significantly reduced. Similar results were revealed for G1 ($t = 2.901$, df = 43, $p < 0.01$) and G4 ($t = 2.643$, df = 43, $p < 0.05$) for trips made to visit relatives or friends. As for the personal business among the participants across groups, only the trips made in the pre-intervention and post-intervention programmes for G3 ($t = 2.606$, df = 43, $p < 0.05$) demonstrated a statistically significant difference.

5.2 Time Allocation for Daily Activities

The time taken for a single trip (in minutes) between two destinations reflects the amount of time allocated for every activity among the participants. In particular, Fig. 5.4 illustrates the average allocated time for every listed activity prior to the implementation of intervention programmes. The participants spent a considerable number of hours in school and for extra academic class during the weekdays, which is rather similar to working adults at their workplace. Specifically, these participants allocated about 413.4 min (6 h 63 min) per person in school and 131.6 min (2 h 11 min) per person for extra academic class on average.

They have mandatory fixed time for school throughout the week, except for weekends. There were approximately 38.6% of total participants who had extra academic class on weekdays, while 30.7% of total participants attended extra academic class over the weekend. However, the time allocation for the extra academic class was higher on the weekends were shown higher with an average of 239.8 min (3 h 59 min) per person, which implies that these participants did spend time on studying over the weekends.

Besides the mandatory school activities and extra academic class, sports achieved the third highest time allocation among the listed activities, with an average of 101.9 min (1 h 41 min) per person on weekdays. There were approximately 10.8% of total participants who participated in sports on weekdays. Over the weekends, the number of participants increased to 22.7% with a higher average of

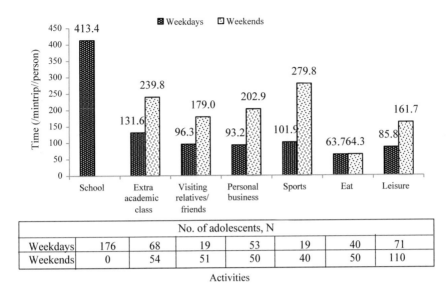

No. of adolescents, N						
Weekdays	176	68	19	53	19	40	71
Weekends	0	54	51	50	40	50	110

Activities

Fig. 5.4 Time allocation for daily activities in pre-intervention programmes

279.8 min (4 h 39 min) per person. The participants demonstrated a higher propensity to take part in the outdoor or recreational activities during the weekends.

Likewise, most of the non-school activities, such as visit relatives or friends, personal business and leisure, received higher participation and longer time allocation from the participants over the weekends. The number of participants who visited relatives or friends on weekends (30.7%) was higher compared to weekdays (10.8%). The time allocation for this social visit increased from an average of 96.3 min (1 h 36 min) per person on weekdays to 96.3 min (2 h 59 min) per person on weekends. As for personal business trips, there were 28.4% of total participants on weekdays (with an average of 93.2 min per person) and 30.1% of total participants on weekends (with an average of 202.9 min per person). In other words, these participants mostly performed their personal business over the weekends since the trips to bank, clinic or hospital, religious place (mosque, church or temple) or any place for running errands cannot be made during school hours on weekdays. As for leisure activities, 40.3% of total participants allocated an average of 85.8 min (1 h 25 min) per person on weekdays whereas 62.5% spent an average of 161.7 min (2 h 41 min) per person over the weekends. Evidently, the participants spent their off-school hours to shop, socialize with friends or catch a movie. Surprisingly, most of these participants prefer to have meals at home, rather than dining out with average time allocation of 63.7 min (1 h 3 min) per person on weekdays and 64.3 min (1 h 4 min) per person on weekends.

Subsequently, Fig. 5.5 illustrates the average allocated time for every listed activity following the implementation of intervention programmes. Given their mandatory fixed time at school, the allocated time spent in school with the

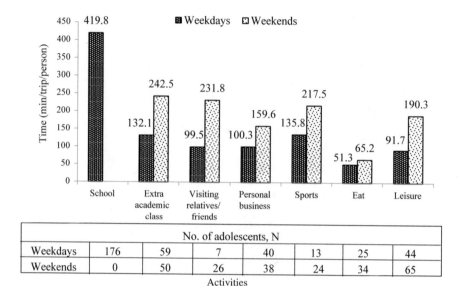

No. of adolescents, N						
Weekdays	176	59	7	40	13	25	44
Weekends	0	50	26	38	24	34	65

Activities

Fig. 5.5 Time allocation for daily activities in post-intervention programmes

implementation of intervention programmes among the participants was almost similar with a slight average increase of 419.8 min (6 h 59 min) per person on weekdays. After school hours, 7.4% of the total participants opted for sports, which contributed an average of 135.8 min (2 h 15 min) per person. There were also other participants who had to attend extra academic class after school hours, which is approximately 132.1 min (2 h 12 min). The participants spent lesser time for personal business (an average of 100.3 min per person), visit relatives or friends (an average of 99.5 min per person), leisure (an average of 91.7 min per person) and dine out (an average of 51 min per person) during the weekdays.

Meanwhile, the participants predominantly spent their time for extra academic class over the weekends with an average of 242 min (4 h 2 min) per person in the post-intervention programmes. Following that, the next common activity gained an average of 231.8 min (3 h 51 min) per person among the participants, while they spent an average of 217.5 min (3 h 37 min) per person for sports. Surprisingly, the participants spent lesser time for leisure (190.3 min per person) with the implementation of intervention programmes. The personal business trip consumed lesser time, resulting in an average of 159.6 min (2 h 39 min) per person. The participants spent the least amount of time on dining out with an average of 65.2 min (1 h 5 min) per person over the weekends, which implied their preference to have meals at home. Overall, the participants attempted to reduce the allocated time for sports, leisure, and personal business with the implementation of intervention programmes.

5.3 Hourly Trip Frequencies

It was apparent that these participants spent most of their time in school on weekdays. Besides that, the participants also participated in various non-school activities throughout the week. Apart from the distribution of average trips per day, the specific number of trips according to the purpose of travel within a specific timeframe was determined based on the obtained travel journal of the participants. This was plausible as the participants recorded their daily travel in time sequence.

Figure 5.6 presents the distribution of average trip frequencies by the hour-interval before the implementation of intervention programmes, which was obtained by dividing the number of trips to the number of days (5 days for weekdays; 2 days for weekends). Meanwhile, Fig. 5.7 presents the distribution of activities by the hour-interval on weekdays and weekends. It should be noted that 00:00 initiated the first hour of the day and 23:00 marked the final hour of the day.

It was revealed that there were four phases of peak hour during the weekdays: (1) 05:00–08:00 h; (2) 13:00–15:00 h; (3) 16:00–20:00 h; (4) 21:00–23:00 h. The first phase of peak hour (05:00–08:00 h) represented the trip to school, which involved all participants across groups, with the highest average hourly trip at 06:00 (93.6 trips). In other words, this implies that there were approximately 93.6 trips to school per day on average. In the second phase of peak hour (13:00–15:00 h), the highest average hourly trip was at 14:00 with 87.6 trips, which indicates the end of school hours and most of the participants travelled back home (389 trips).

As shown in Fig. 5.7a, only certain participants pursued other non-school activities after school hours (39 trips). As for the third phase of peak hour (16:00–20:00 h), the distribution of these average hourly trips fluctuated, in which the

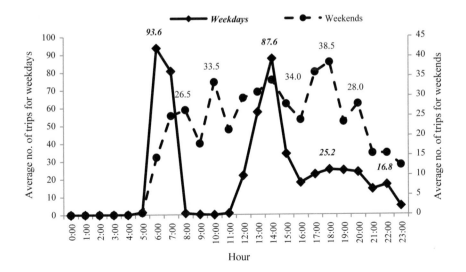

Fig. 5.6 Distribution of average hourly trips in pre-intervention programmes

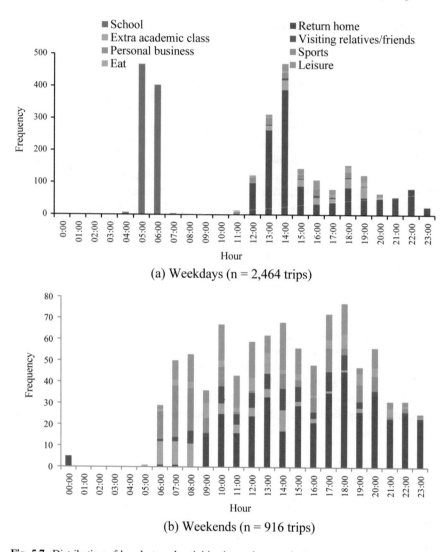

Fig. 5.7 Distribution of hourly travel activities in pre-intervention programmes

highest average hourly trip was at 18:00 with 25.2 trips. This revealed that most trips at 18:00 involved returning home (85 trips) despite the various non-school activities at various timeframes. The final phase of peak hour (21:00–23:00 h) recorded the highest average hourly trip of 16.8 trips at 22:00, which may imply that the participants may have just completed their extra academic class or dined out before returning home. There were higher fluctuations in the average hourly trips over the weekends since the participants were not confined by the mandatory school hours. The number of trips made over the weekends amounted to 916 with five phases of peak hour: (1) 07:00–09:00 h; (2) 09:00–11:00 h; (3) 13:00–15:00 h; (4) 17:00–19:00 h; (5) 19:00–21:00 h.

Referring to Fig. 5.7b, the first phase of peak hour (07:00–09:00) revealed the highest average hourly trip (26.5 trips) at 08:00, which was mainly occupied by leisure (13 trips), extra academic class (11 trips), and sports (11 trips) activities. There were lesser trips for personal business, visiting relatives or friends, or having breakfast away from home in the morning. Subsequently, the next phase of peak hour (09:00–11:00 h) revealed that most of the participants travelled back home (25 trips) from their prior activities, followed by trips to other activities, such as leisure (17 trips), visiting relatives or friends (8 trips), personal business (7 trips), extra academic class (5 trips), dine out (3 trips), and sports (2 trips). The highest average hourly trip (33.5 trips) for this phase occurred at 10:00. A similar pattern was revealed in the subsequent hours with the highest average hourly trip at 38.5 trips at 18:00, which was predominantly occupied by trips to home (45 trips). Overall, the participants exhibited a significant number of simple trip chains in the pre-intervention programmes.

On the other hand, Fig. 5.8 presents the distribution of average trip frequencies by the hour-interval with the implementation of intervention programmes, while Fig. 5.9 reveals the distribution of activities by the hour-interval on weekdays and weekends in support of Fig. 5.8. Similarly, there were also four identified phases of peak hour during the weekdays. The first phase of peak hour (05:00–07:00 h) revealed the highest average of the hourly trip (95.0 trips) at 06:00 with a total of 475 trips to school given their mandatory fixed time in school daily on weekdays.

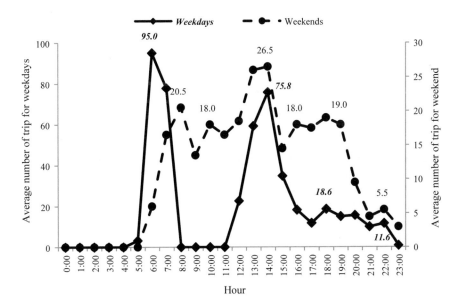

Fig. 5.8 Distribution of average hourly trips in post-intervention programmes

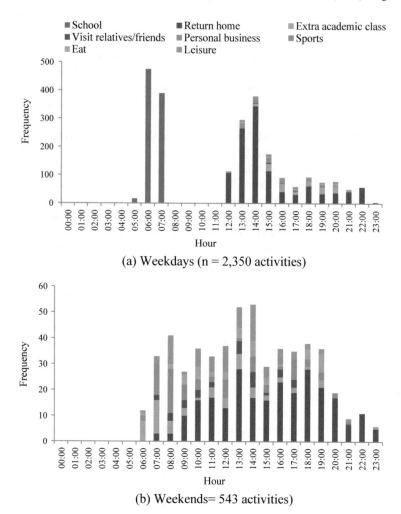

Fig. 5.9 Distribution of hourly travel activities in post-intervention programmes

After regular school hours, the participants took part in various trips, which contributed to various trip chaining patterns. This then led to the second phase of peak hour (13:00–15:00 h) with the highest average hourly trip (75.8 trips) at 14:00, which marked the end of school hours.

Most of the participants travelled back home (344 trips), while the remaining 35 trips made were associated to personal business (14 trips), extra academic class (9 trips), visit relatives or friends (2 trips), and sports (1 trip). The third phase of peak hour (16:00–21:00 h) was predominantly occupied by trips to return home (61 trips) after their non-school activities among the participants. The highest average hourly trip was at 18:00 with 18.6 trips. Last but not least, most of the participants were likely to travel back home after their extra academic class in the final phase of peak hour (21:00–23:00 h) during the weekdays in the post-intervention programmes. In particular, the highest average hourly trip was recorded at 22:00 with 11.6 trips.

Likewise, there were also apparent fluctuations in the average hourly trips over the weekends given the off-school hours. As shown in Fig. 5.9b, there were six phases of peak hours identified over the weekends following the implementation of intervention programmes: (1) 07:00–09:00 h; (2) 09:00–11:00 h; (3) 13:00–15:00 h; (4) 15:00–17:00 h; (5) 17:00–19:00 h; (6) 21:00–23:00 h. Most of the activities in the initial peak hours over the weekends were dominated by trips for leisure (11 trips), personal business (9 trips), sports (8 trips), and extra academic class (5 trips). There were lesser trips to visit relatives or friends (3 trips), return home (3 trips), and have breakfast at the café or restaurant (2 trips). The highest average hourly trip was recorded at 08:00 with 20.5 trips. Following that, the second phase of peak hours was initiated at 09:00 with the highest average hourly trip at 10:00 (33.5 trips), which was dominated by trips to return home (16 trips) after the non-school activities. Undeniably, there were participants who also travelled for leisure (7 trips), dining out (5 trips), visiting relatives or friends (3 trips), extra academic class (1 trip) and sports (1 trip).

There were similar patterns with trips to return home as the main activity with the passing of time, which extended to the next two phases of peak hours after 13:00–15:00 h. The highest average hourly trip (18.0 trips) occurred at 16:00 in the fourth phase of peak hours (15.00–17.00 h) and another at 18:00 in the following phase of peak hours (17:00–19:00 h) with 19.0 trips. The prior phase of 13:00–15:00 h depicted the highest average of the hourly trip (26.5 trips) at 14:00 among the average of hourly trips per day among the participants over the weekends. In particular, it was dominated by the participants' trips for leisure (14 trips), dining out (6 trips), and visiting relatives or friends (6 trips). There were lesser trips for personal business (4 trips), sports (2 trips), and extra academic class (1 trip). Last but not least, the final phase of peak hours (21:00–23:00 h) recorded an average of the hourly trip of 5.5 trips at 22:00, in which most of the participants travelled back home (11 trips) after the prior weekend trips for non-school activities.

5.4 Choice of Transport for Travel

Figure 5.10 presents the choice of transport for travel with respect to the purpose of travel prior to the implementation of intervention programmes among the participants. Majority of the participants opted for private vehicles for their daily travel during the weekdays. For instance, 40.2% of the total numbers of trips to school (880 trips) were by car, while 29.0% of these trips were via motorcycle. However, there were lesser trips via bus (18.8%) and van (3.4%). Besides that, the participants attempted to walk (8.2%) and cycle (0.5%) to school. Apart from school as destination, the participants also heavily depended on cars for dining out (62.9%),

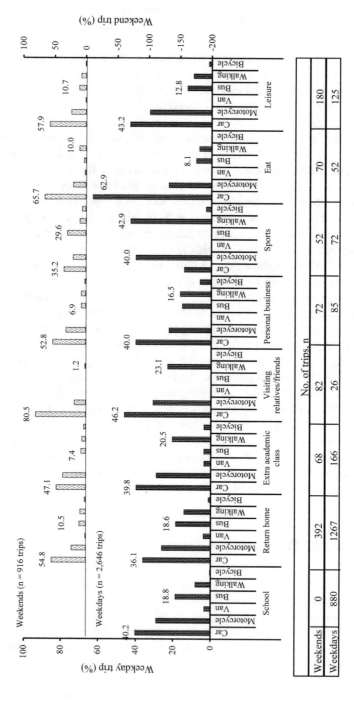

Fig. 5.10 Choice of transport for travel according to the purpose of travel in pre-intervention programmes

visiting relatives or friends (46.2%), personal business (40.0%), extra academic class (39.8%), returning home (36.1%) and leisure (33.2%). Unexpectedly, the participants opted to walk when it comes to sports trips. This may be plausible when their destinations are located within walking distance. Additionally, McDonald (2005) propounded the significance of walking as part of their sports activities.

The choice of transport for travel over the weekends demonstrated a similar pattern, in which cars and motorcycles were mainly opted for the majority of the trips made by the participants. Cars were mainly selected as the choice of transport for travel (>50.0%) to return home, visit relatives or friends, personal business, dine out, and for leisure. Adding to that, cars were also extensively used as the preferred form of transport among the participants to travel for extra academic class (47.1%) and sports (35.2%). The extensive use of private vehicles over the weekends indicates the plausibility that the travel behaviour of participants was rather affected by the household activities and the extensive travel distance (in which independent travelling among adolescents was not encouraged). For instance, it was revealed that most of the trips to visit relatives or friends (80.5%) were via car.

In general, the participants demonstrated their reliance on private vehicles for their daily travel based on the number of trips made via cars and motorcycles prior to the implementation of intervention programmes. Despite that, there were several participants who attempted to travel by public transports or non-motorized transports (such as cycling and walking), which were commendable given the exclusion of intervention programmes at this point.

Following the implementation of intervention programmes among the participants, the choice of transport for travel with respect to the purpose of travel was subsequently assessed. Referring to Fig. 5.11, the results revealed that the preferred choice of transport for travel was not exclusively private vehicles, particularly in their return trips. These participants demonstrated their preference to travel by car (25.3%) and motorcycle (33.6%) to an extra academic class given the need of punctuality to arrive at their destination. However, these private vehicles were not necessarily opted for when they had to travel back home after the extra academic class. Although cars (35.0%) and motorcycles (26.3%) dominated as the choice of transports for travel among the participants, especially to school during the weekdays, the implementation of intervention programmes prompted the participants to support and commit to lower carbon mobility, as reflected by the increase of bus usage (25.9%) among the participants.

Apart from school, most trips, except for extra academic class, sports, and dining out, demonstrated a substantial increase in the bus usage. With the implementation of intervention programmes, the participants' preferred choice of transport to travel back home was evidently the bus with 458 trips. Nevertheless, there was lesser bus usage among participants who travelled for leisure (21 trips), personal business (19 trips), sports (6 trips) and visiting relatives or friends (5 trips). Besides that, a total of 10.7% of school trips were involved walking and cycling, only 2.2% of trips involved private van. Surprisingly, walking trips for various purposes, such as sports (48.0%), dining out (42.4%) and personal business (33.9%), were rather

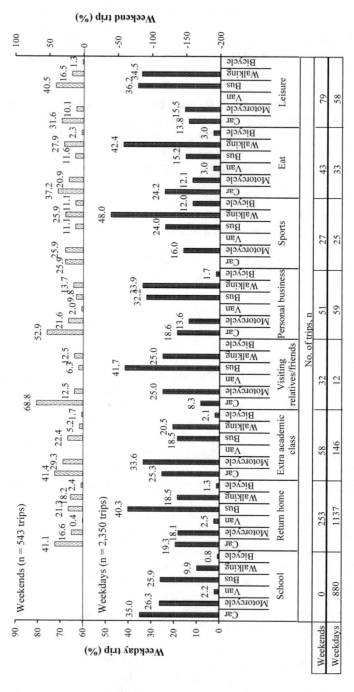

Fig. 5.11 Choice of transport for travel according to the purpose of travel in post-intervention programmes

apparent in the post-intervention programmes among the participants, which indicates the possibility of walking distance between destinations after all since public transport service may consume more time.

Despite the high usage of private vehicles to travel for non-school activities, there were lesser trips made over the weekends following the implementation of intervention programmes. The participants were rather dependent on private vehicles based on the number of trips made via car for visiting relatives or friends (81.3%), personal business (74.5%), sports (74.5%), extra academic class (57.7%) and dining out (58.1%). Nonetheless, the bus usage for leisure trips was high, which implies the plausibility of group travel with their peers. Meanwhile, walking was opted as an alternative to travel for dining out (27.9%) and sports (25.9%), which reaffirms the possibility of walking distance between destinations.

Accordingly, Fig. 5.12 depicts the average number of trips in the choice of transport for travel across groups in the pre-intervention and post-intervention programmes among the participants, while Table 5.3 tabulates the results of paired-sample t-test between the pre-intervention and post-intervention programmes across groups. First, the average number of trips via car as the choice of transport for travel statistically significantly differed between the pre-intervention and post-intervention programmes for G1 ($t = 2.453$, df = 43, $p < 0.05$), G2 ($t = 3.587$, df = 43, $p < 0.001$), G3 ($t = 6.286$, df = 43, $p < 0.001$) and G4 ($t = 4.430$, df = 43, $p < 0.001$) during the weekdays. In particular, G3 demonstrated the highest mean difference ($M_{\text{diff}} = 4.136$, SD = 4.365) among these intervention groups, which indicated the highest reduction in the car usage among the participants in this particular group.

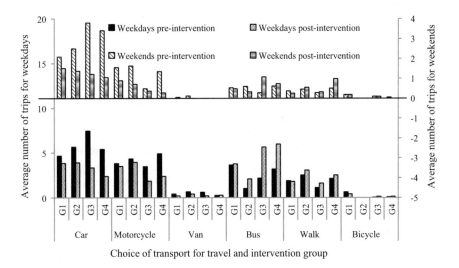

Fig. 5.12 Average number of trips on weekdays and weekends for pre-intervention and post-intervention programmes

Table 5.3 Paired-sample t-test results of choice of transport for travel according to intervention groups

Choice of transport for travel	G1		G2		G3		G4	
	M_{diff}	t-value	M_{diff}	t-value	M_{diff}	t-value	M_{diff}	t-value
Weekday travel								
Car	**0.82**	**2.45**	**1.75**	**3.59**	**4.14**	**6.29**	**3.02**	**4.43**
Motorcycle	0.32	1.09	0.39	1.32	**1.64**	**3.68**	2.52	3.23
Van	0.22	1.93	0.25	1.67	0.41	1.62	−0.02	−0.44
Bus	−0.14	−0.97	**−1.07**	**−2.98**	**−3.46**	**−7.51**	**−2.80**	**−5.01**
Walk	0.09	0.37	−0.54	−1.56	−0.50	−1.12	−0.39	−1.16
Bicycle	0.23	1.20	–	–	−0.09	−0.81	−0.05	−0.44
Weekend travel								
Car	**0.59**	**2.22**	**1.14**	**3.22**	**2.59**	**4.91**	**2.37**	**5.66**
Motorcycle	**0.66**	**2.20**	**0.93**	**2.68**	0.11	0.61	**1.07**	**3.14**
Van	−0.05	−1.00	0.11	1.40	–	–	–	–
Bus	0.05	0.31	0.27	1.23	**−0.80**	**−2.81**	−0.11	−0.55
Walk	0.11	1.00	−0.10	−0.43	−0.05	−0.31	−0.48	−1.80
Bicycle	0.00	0.00	–	–	0.00	0.000	0.05	1.00

Notes Highlighted data in bold denotes statistical significance at 0.05 level; *G1* represents the control group; *G2* represents the journal planning group; *G3* represents the incentive group; *G4* represents the journal planning with an incentive group

This was followed by G4 ($M_{diff} = 3.023$, SD $= 4.526$), G2 ($M_{diff} = 1.750$, SD $= 3.236$), and lastly, G1 ($M_{diff} = 0.818$, SD $= 2.213$). Meanwhile, G4 ($t = 3.229$, df $= 43$, $p < 0.05$) recorded the highest significant mean difference for the average number of trips via motorcycle as the choice of transport for travel during the weekdays, which was followed by G3 ($t = 3.680$, df $= 43$, $p < 0.05$). In other words, the implementation of intervention programmes depicted the effectiveness of combining both journey planning strategy and distribution of free bus pass or any other forms of incentive in significantly reducing the participants' dependence on motorcycle in their daily travel. Besides that, the mean difference in the number of trips via bus as the choice of transport for travel was not significant between the pre-intervention and post-intervention programmes in G1 only. Accordingly, G3 ($t = -7.505$, df $= 43$, $p < 0.001$) recorded the highest significant mean difference. Another significant observation was that the average number of trips by bus for G4 ($t = -5.010$, df $= 43$, $p < 0.001$) surpassed the average number of trips by bus for G2 ($t = -2.982$, df $= 43$, $p < 0.05$) in the case of weekday travel. In this case, the negative sign in the mean difference indicated the increase in bus usage in the post-intervention programmes.

Adding to that, the trend in the average number of trips in the choice of transport for travel over the weekends was similarly portrayed. Overall, the participants seemed to reduce the number of trips via private vehicles and demonstrated a higher dependence on sustainable transports for their weekend travel with the implementation of intervention programmes. For instance, the implementation of intervention programmes effectively reduced the number of trips via car across groups, in which G3 attained the highest significant mean difference (M_{diff} = 2.591, t = 5.568, p > 0.05) and followed by G4 (M_{diff} = 2.364, t = 4.911, p > 0.05). Similar results for the use of motorcycle for travel were observed among the participants in G4 (M_{diff} = 1.068, t = 3.141, p > 0.05), G2 (M_{diff} = 0.932, t = 2.680, p > 0.05), and lastly, G1 (M_{diff} = 0.659, t = 2.200, p > 0.05).

As opposed to the weekday travel, there was only one intervention group (G3: M_{diff} = −0.795, t = −2.814, p < 0.05) that demonstrated the significant mean difference in the number of bus usages over the weekends instead, which depicts significant increase following the implementation of intervention programmes. Besides that, the implementation of intervention programmes did not significantly affect the choice of transport for travel when the participants had to travel by foot, bicycle, or van during the weekdays, which demonstrates the plausibility of commitment among the participants towards travelling via sustainable transports (e.g. bus) in reducing their carbon footprint. However, the effects of intervention programmes towards the choice of transport for travel among the participants were demonstrated over the weekends, as reflected by the reduction in the number of trips via private vehicles to travel in the post-intervention programmes.

5.5 Travel Distance and Travel Duration

Essentially, the calculation of carbon emissions requires data on the travel distance. Adding to that, the travel durations may differ despite similar travel distance. There are several factors that may contribute to such circumstances, namely the choice of transport for travel, the traffic flow, the purpose of travel, and the personality of the driver. Accordingly, Fig. 5.13 presents the average travel distance as well as the travel duration among the participants with respect to their choice of transport for travel in the pre-intervention programmes.

Travelling by car recorded the highest average trip distance (4.3 km/trip/person) with a longer average travel duration of 13.7 min/trip/person during the weekdays. Although travelling by bus recorded lower average trip distance (2.8 km/trip/person) and average travel duration (12.0 min/trip/person) compared to travelling by car, travelling by bus seemed to consume more time considering the distance travelled in this case. Meanwhile, the average trip distance for those who travelled via motorcycle recorded 1.7 km/trip/person within an average travel duration of 6.4 min/trip/person. In other words, those who opted for public transports, such as bus, used up more time on travelling for the same travel distance via private vehicles.

Likewise, the weekend travel also displayed similar observations in terms of travel distance and travel duration. However, the results revealed the propensity of these participants to cover longer travel distance and travel duration over the weekends. In particular, the participants travelled an average of 5.3 km/trip/person

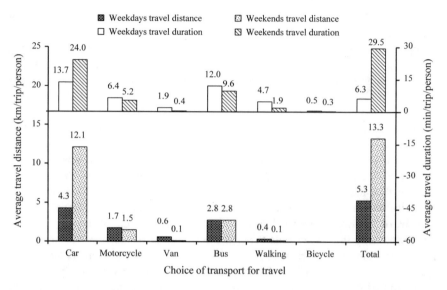

Fig. 5.13 Average travel distance and travel duration for pre-intervention programmes

within 6.3 min/trip/person during the weekdays, while their weekend travel recorded an average of 13.3 km/trip/person within 29.5 min/trip/person. Accordingly, the average trip distance for travelling by car increased to 12.1 km/trip/person and the corresponding average trip distance extended to 24.0 min/trip/person. However, the participants recorded shorter average travel duration (9.6 min/trip/person) despite similar average travel distance (2.8 km/trip/person). The disparity in the travel duration may be caused by the drop-off and pick-up of other bus passengers before the participants arrived at their destination. Another plausible rationalization for this scenario was that the participants travelled by bus during off-peak hours.

Following the implementation of intervention programmes, Fig. 5.14 revealed that the participants generally travelled lesser in terms of travel distance during the weekdays, specifically an average of 5.1 km/trip/person. However, their overall average travel duration was extended to 21.0 min/trip/person. Similarly, the overall weekend travel recorded lower average travel distance of 5.5 km/trip/person within an average of 18.3 min/trip/person.

Accordingly, a bus trip for weekday travel recorded the highest average travel distance of 4.2 km/trip/person within an average travel duration of 19.4 min/trip/person, which was reportedly similar to the recorded travel distance for a car trip in the pre-intervention programmes. However, the bus trip in the pre-intervention programmes recorded higher average travel duration. As for the weekend travel by bus, an average travel duration of 1.8 km/trip/person was recorded within the highest average travel duration of 8.1 min/trip/person.

On the other hand, the average travel distance of a car trip for weekday travel recorded 3.1 km/trip/person within an average travel duration of 10.1 min/trip/

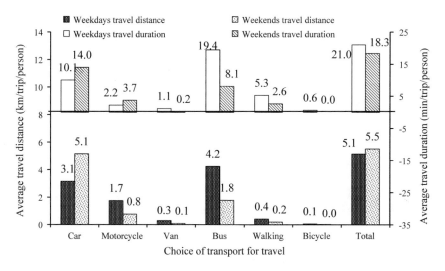

Fig. 5.14 Average travel distance and travel duration for post-intervention programmes

person, while the weekend travel by car recorded an average travel distance of 5.1 km/trip/person within an average travel duration of 14.0 min/trip/person. The participants seemed to spend lesser time in their daily travel via car compared to their daily travel via bus, which reaffirms the implications of drop-off and pick-up of other bus passengers before the participants arrived at their destination.

Meanwhile, travelling via motorcycle recorded an average travel distance of 1.7 km/trip/person within an average travel duration of 2.2 min/trip/person during the weekdays. However, the participants demonstrated a shorter average travel distance of 0.8 km/trip/person, but recorded a longer average travel duration of 3.7 min/trip/person. The average travel duration for those who travelled by foot recorded 5.3 min/trip/person for a mere average distance of 0.4 km during the weekdays, which were subsequently reduced over the weekends in both average travel distance (0.2 km/trip/person) and average travel duration (2.6 min/trip/ person).

Following that, Fig. 5.15 depicts the average travel distance in the choice of transport for travel across groups in the pre-intervention and post-intervention programmes among the participants while Table 5.4 tabulates the results of paired-sample *t*-test in the travel distance between pre-intervention and post-intervention programmes across groups. The overall travel distance for trips via car, motorcycle and van during the weekdays with the implementation of intervention programmes were reportedly reduced across groups. On the contrary, the overall travel distance for trips by foot, bicycle and bus were reportedly increased across groups instead. These reported results demonstrated the effectiveness of intervention programmes in promoting the use of sustainable transports and reducing the need to travel via private vehicles, which subsequently reduced the carbon emissions from these vehicles.

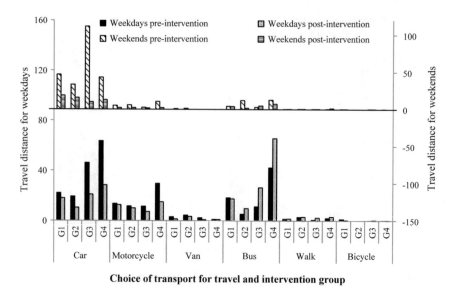

Fig. 5.15 Average travel distance for pre-intervention and post-intervention programmes

Table 5.4 Paired-sample *t*-test results of travel distance in the choice of transport for travel according to intervention groups

Choice of transport for travel	G1		G2		G3		G4	
	M_{diff}	t-value	M_{diff}	t-value	M_{diff}	t-value	M_{diff}	t-value
Weekday travel								
Car	4.13	1.84	**8.82**	**3.24**	**25.18**	**5.21**	**35.23**	**4.99**
Motorcycle	0.99	0.70	1.61	0.97	**4.19**	**3.12**	**14.62**	**2.25**
Van	1.71	1.52	1.05	1.45	1.56	1.63	0.09	0.34
Bus	0.79	0.58	**−4.43**	**−2.67**	**−15.14**	**−6.13**	**−23.48**	**−3.81**
Walk	−0.04	−0.10	−0.17	−0.75	**−1.67**	**−2.38**	−0.88	−1.13
Bicycle	0.88	1.07	–	–	−0.39	−1.27	−0.01	−0.24
Weekend travel								
Car	**28.34**	**2.03**	1.06	1.06	**101.58**	**2.86**	**30.43**	**3.94**
Motorcycle	2.87	1.78	3.55	1.98	0.87	1.30	**8.46**	**2.33**
Van	−0.77	−1.00	1.18	1.21	–	–	–	–
Bus	0.25	0.24	**10.49**	**2.14**	−1.80	−0.86	5.36	0.75
Walk	−0.05	−0.35	0.04	0.26	0.09	0.40	−0.93	−1.82
Bicycle	0.01	0.24	–	–	0.02	0.27	0.05	1.00

Notes Highlighted data in bold denotes statistical significance at 0.05 level; *G1* represents the control group; *G2* represents the journal planning group; *G3* represents the incentive group; *G4* represents the journal planning with an incentive group

When it concerns travelling by car during the weekdays, G4 (t = 4.997, df = 43, p < 0.05) attained the largest significant mean difference (M_{diff} = 35.225, SD = 46.755), which indicated the highest reduction in the travel distance following the implementation of intervention programmes among the participants. Subsequently, both G3 (t = 5.212, p < 0.05) and G2 (t = 3.235, p < 0.05) also demonstrated significant mean differences (G3: M_{diff} = 25.180, SD = 32.048; G2: M_{diff} = 8.823, SD = 18.088) in the travel distance. Besides that, the participants in G4 who travelled via motorcycle also displayed the highest reduction in the travel distance (M_{diff} = 14.623, SD = 43.084) with significant mean difference (t = 2.251, df = 43, p < 0.05), which was followed by G3 (M_{diff} = 4.191, SD = 8.911; t = 3.120, df = 43, p < 0.05). Only both of these intervention groups, with the exception of G1 and G2, exhibited a significant mean difference of the travel distance in the weekday travel via motorcycle between the pre-intervention and post-intervention programmes. These results reaffirm the effectiveness of combining various intervention techniques in ensuring the adolescents to reduce their daily travel via motorcycle.

Despite the significant reduction in the travel distance via private vehicles (of both car and motorcycle), G4 (t = −3.814, df = 43, p < 0.001), G3 (t = −6.126, df = 43, p < 0.01), and G2 (t = −2.666, df = 43, p < 0.05) demonstrated significant increase in the travel distance via bus, as denoted by the negative difference in t-value. In other words, the travel distance for the weekday travel by bus prior to the implementation of intervention programmes was shorter. Specifically, the largest mean difference of the travel distance for a bus trip between the pre-intervention and post-intervention programmes was achieved by G4 (M = −23.480, SD = 40.834), which was followed by G3 (M = −15.136, SD = 16.390) and lastly, G2 (M = −4.432, SD = 11.026).

Following the implementation of intervention programmes, the travel distance for trips via car and motorcycle also significantly decreased across groups over the weekends. Despite the reduction in the travel distance for trips via car, the mean difference for G2 was not significant. The participants in G3 (t = 2.858, df = 43, p < 0.05) demonstrated the largest significant reduction in the travel distance (M_{diff} = 101.575, SD = 235.715), which was followed by G4 (M_{diff} = 30.427, SD = 51.251; t = 3.938, p < 0.001) and lastly, G1 (M_{diff} = 28.355, SD = 92.888; t = 2.025, p < 0.05). These results imply the plausibility of unreported extra travel distance in possible family or friend trips among the participants throughout the intervention period. As for the trips involving motorcycle, only G4 reported the significant mean difference in the travel distance (M_{diff} = 8.455; t = 2.332, p < 0.05).

However, there were only a slight increase in the travel distance for trips via bus and travelling by foot. Considering that only G2 with significant reduction in the travel distance for trips via bus (M_{diff} = 10.489; t = −2.136, p < 0.05), it was revealed that the participants did attempt to commit to the intervention programmes by making use of the public transports for their daily travel during the weekdays particularly, but the participants demonstrated considerable dependence on private vehicles over the weekends. Thus, the travel distance in the choice of transport for travel was almost similar to the changes of mode choice frequencies, which implies the significant effects of the choice of transport for travel towards the overall travel distance among the participants.

In terms of travel duration, Fig. 5.16 shows the average duration for pre- and post-intervention based on different mode of transport. Meanwhile, Table 5.5 presents the paired-sample t-test results based on intervention groups. Likewise, G4 (M_{diff} = 50.909, SD = 113.703; t = 2.970, df = 43, p < 0.05) again reported the highest significant reduction in the travel duration for trips via motorcycle. The participants in G3 (M_{diff} = 26.477, SD = 49.601; t = 3.541, df = 43, p < 0.05) also demonstrated a significant reduction in their travel duration when they travelled via motorcycle during the weekdays. However, their weekend trip via motorcycle revealed <u>contradictory observations, which may be contributed by inaccurate</u> reporting of travel duration or traffic congestion. Although the participants in G4 demonstrated a significant reduction in the travel distance over the weekends, their travel duration was not significantly reduced. In fact, only G1 (M_{diff} = 8.341, SD = 36.141; t = 2.140, df = 43, p < 0.05) and G2 (M_{diff} = 12.182, SD = 36.415; t = 2.219, df = 43, p < 0.001) reported statistically significant reduction in the travel duration for weekend travel via motorcycle.

As for those who travelled by bus during the weekdays, G3 attained the highest significant increase in the travel duration (M_{diff} = −95.159, SD = 103.249; t = −6.113, df = 43, p < 0.05) with the implementation of intervention programmes among the participants. Expectedly, other groups, namely G2 (M_{diff} = −35.636, SD = 78.245; t = −3.021, df = 43, p < 0.05) and G4 (M_{diff} = −79.500, SD = 149.910; t = −63.518, df = 43, p < 0.001) also reported a

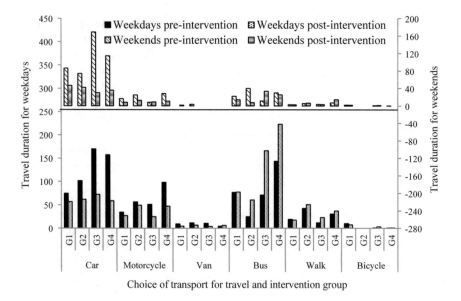

Fig. 5.16 Average travel duration for pre-intervention and post-intervention programmes

Table 5.5 Paired-sample t-test results of travel duration in the choice of transport for travel according to intervention groups

Choice of transport for travel	G1		G2		G3		G4	
	M_{diff}	t-value	M_{diff}	t-value	M_{diff}	t-value	M_{diff}	t-value
Weekday travel								
Car	17.89	1.97	**39.72**	**3.50**	**97.71**	**5.20**	**98.89**	**4.70**
Motorcycle	7.068	1.30	6.98	1.00	**26.48**	**3.54**	**50.91**	**2.97**
Van	4.773	1.47	4.77	1.69	6.61	1.36	−1.93	−1.04
Bus	−0.773	−0.09	**−35.64**	**−3.02**	**−95.16**	**−6.11**	**−79.50**	**−3.52**
Walk	1.591	0.56	−7.96	−1.30	−11.34	−1.87	−6.43	−0.82
Bicycle	2.773	1.32	–	–	−2.50	−0.88	−0.11	−0.28
Weekend travel								
Car	**39.46**	**2.19**	31.55	1.51	**138.61**	**3.49**	**78.41**	**4.48**
Motorcycle	**8.34**	**2.14**	**12.18**	**2.22**	−0.80	−0.21	17.36	1.88
Van	−1.71	−1.00	3.61	1.42	–	–	–	–
Bus	7.66	1.42	**32.59**	**2.03**	**−22.00**	**−2.22**	4.48	0.39
Walk	0.25	0.14	−1.09	−0.48	0.43	0.18	−7.27	−1.44
Bicycle	0.34	0.55	–	–	−0.46	−0.63	0.25	1.00

Notes Highlighted data in bold denotes statistical significance at 0.05 level; *G1* represents the control group; *G2* represents the journal planning group; *G3* represents the incentive group; *G4* represents the journal planning with an incentive group

significant increase in the travel duration. However, when it comes to weekend travel, only G2 ($t = 2.025$, df = 43, $p < 0.05$) and G3 ($t = -2.218$, df = 43, $p < 0.05$) reported significant mean difference in the travel duration, but with contradictory observation—the participants in G2 demonstrated significant reduction ($M_{\text{diff}} = 32.591$, SD = 106.755) whereas those in G3 exhibited significant increase ($M_{\text{diff}} = -22.000$, SD = 65.785). In short, the obtained results of travel distance and travel duration in the choice of transport for travel revealed that the two aspects were aligned with the propensity to reduce the use of private vehicles in terms of travel duration by extending the travel duration via bus among the participants.

5.6 Carbon Emissions

Meanwhile, Fig. 5.17 presents the results on the carbon emissions from vehicles, specifically in the emission of carbon dioxide (CO_2), across groups in the pre-intervention and post-intervention programmes for the weekday travel and weekend travel among the participants. Prior to the implementation of intervention programmes, the participants produced between 89.4 and 265.4 kg of CO_2 in their weekday travel in the following ascending order: G1 < G2 < G3 < G4.

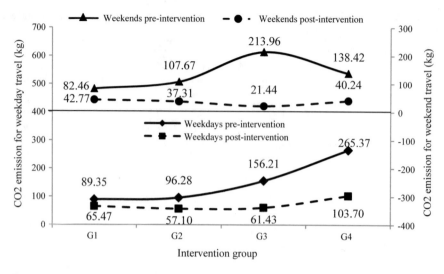

Fig. 5.17 Carbon emissions from vehicles on weekdays and weekends

Considering that the travel distance and the choice of transport for travel are important inputs for the calculation of carbon emissions, it was revealed that the participants in G4 produced the highest amount of CO_2 during the weekdays. In other words, they had more trips or longer travel distance with the use of private vehicles, such as car and motorcycle, compared to the participants in other groups. However, the amount of CO_2 emission for weekend travel prior to the implementation of intervention programmes was lower, which ranged between 82.5 and 214.0 kg of CO_2 in the following ascending order with G3 recording the highest amount of CO_2 emission: G1 < G2 < G4 < G3. When it comes to family trip over the weekend, the participants may not have the final decision in the choice of transport for travel with the car as the primary form of transport; thus, resulting to higher carbon emission.

Following the implementation of intervention programmes, the amount of carbon emission was recorded between 57.1 and 103.7 kg of CO_2 for the weekday travel. The participants in G4 remained the highest in their amount of carbon emission, which was followed by G1 (65.47 kg), G3 (61.43 kg) and lastly G2 (57.10 kg). The reduction in the amount of carbon emission for G2 was not as substantial as other groups. As for their weekend travel, the amount of carbon emission was between 21.4 and 42.8 kg of CO_2 in the following order: G3 < G2 < G4 < G1. Surprisingly, the participants in G1 produced the highest amount of CO_2 with the smallest observed difference across groups despite recording the lowest amount of carbon emission before the implementation of intervention programmes. Nevertheless, the overall amount of carbon emission remained evidently reduced following the implementation of intervention programmes with the lower number of trips, travel distance and mode change.

5.7 Overall Implications of TBM Programme

The current chapter presents the overall implications of TBM programme, specifically the implementation of various intervention programmes, in assessing the travel behaviour among the participants. With the goal of promoting awareness on low carbon mobility among the participants, G1 received only motivation intervention; G2 received both motivation and journey planning interventions; G3 received both motivation and incentive interventions; G4 received similar interventions with G2 and G3.

Overall, the number of trips for leisure and returning home were clearly reduced during the weekdays, which demonstrate the participants' propensity to reduce single trips with the reduction of non-school activities after mandatory school hours or the extended duration of their activities. The participants in G4 significantly reduced their trips for personal business, returning home, dining out, and leisure, which demonstrates their attempt to make use of the distributed free bus pass throughout the implementation of intervention programmes. Similarly, there were also a significant reduction in the number of trips for returning home, leisure and visiting relatives or friends among the participants in G3. With the free bus pass, the participants were required to plan their daily travel according to the provided bus schedule. Thus, it was inevitable that there may be certain non-school activities that the participants were not able to perform on a usual basis. Apart from that, the motivation alone propelled the participants in G1 to significantly reduce their trips for leisure and returning home in reducing the carbon emission from vehicles. However, the implementation of both motivation and journey planning interventions did not contribute any significant effects on the participants' daily travel given the insignificant differences found in the pre-intervention and post-intervention programmes.

In the case of the purpose of travel, the weekend travel behaviour among the participants in the pre-intervention and post-intervention programmes was generally rather inconsistent despite the substantial significant reduction in the number of trips for non-school activities. The issue of whether the purpose of travel was performed based on an individual decision or household decision was not explicitly specified. For instance, a trip for household grocery shopping and a trip to purchase school items were similarly categorised for the purpose of travel, namely shopping trip. The findings also parallel with previous studies that show the highest trips during weekends were dominated by more leisure-related activities including shopping and socializing (Song et al. 2012; Zhong et al. 2008). Weekends trips can be also involving with other social trips such as going to the house of worship and/ or for wedding ceremony (Agarwal 2004). According to Zhong et al. (2008), the number of non-obligatory trips was distinguished only by 21.5% of the total trips during the weekdays, but the frequency was heightening up to 33% on the weekends.

Furthermore, in this study, the participants in G2, G3 and G4 significantly increased their bus usage (and reduced their car usage) for their daily travel during the weekdays. This reaffirms the positive effects of combining various intervention strategies in encouraging the adolescents to opt for sustainable transports. Despite the significant reduction in the car usage, there was no significant increase in the bus usage among the participants in G1, which reaffirms the inadequacy of providing motivation alone in prompting these participants to change their travel behaviour. Another plausible rationalization for the obtained results on G1 was that most of the participants already opted for public transport in their daily travel prior to the implementation of intervention programmes. Apart from cars, the participants across groups also reported a significant reduction in motorcycle usage over the weekends. Adding to that, only G3 demonstrated a significant shift in the choice of transport for travel. The participants in G3 who received free bus pass were likely to extend their bus usage to their weekend travel due to their positive experience in taking the public transports to move about during the weekdays.

This is contrary to the previous studies, which reported that many people have a tendency to use more private vehicles compared to public transport options during weekends (Koo and Kim 2004). Besides, Jang and Lee (2010), Hunt et al. (2005), Agarwal (2004), and Yai et al. (1995) also revealed that the weekends' trip length and travel duration are usually longer than weekdays' trips. In this case study, despite the significant increase in the travel distance and travel duration, the participants in G2, G3, and G4 reported a significant decrease in the car usage during the weekdays following the implementation of intervention programmes. Nevertheless, despite the effects of travel distance towards travel duration, there was no significant difference reported in G1 for their usage of private vehicles and public transport (i.e. bus). The obtained results in the case of travel distance and travel duration demonstrated a similar pattern for the obtained results in the case of the choice of transport for travel among the participants. Accordingly, the participants in G3 reported the highest significant reduction in the travel distance for their trips via private vehicles, which was followed by G4 and lastly, G1. Notably, the participants' travel behaviour was also potentially affected by the household travel decision, especially for trips to visit relatives. However, there was no significant increase in the travel distance for the alternative transport options over the weekends across groups, which revealed the possibility of inaccurate reporting of daily travel among the participants.

Last but not least, the implementation of intervention programmes also significantly reduced the amount of carbon emission from vehicles during the weekdays, which were in line with the obtained results on the travel distance and the choice of transport for travel among the participants. Accordingly, the use of private vehicles over an extensive travel distance contributes to higher carbon emission. The implementation of various intervention strategies contributed the highest significant difference in the reduction of carbon emission for G4, which was followed by G3, G2 and lastly, G1. Irrefutably, the obtained results reaffirmed the effectiveness of various intervention strategies in inducing sustainable travel behaviour among the participants. The significant reduction in the travel distance for trips via car and the significant increase in the travel distance for trips via bus prompted the participants to demonstrate their effort to reduce their carbon footprint. Interestingly, the amount of carbon emission for G1 was reportedly the lowest with the insignificant mean difference compared to other groups. The implementation of intervention programmes seemed to produce no significant differences in the amount of carbon emissions among the participants in G1, which may be due to their existing sustainable travel behaviour.

As for the weekend travel, the participants in G3 reported a significant reduction in the amount of carbon emission following the implementation of intervention programmes, which was followed by G4 and G1. Meanwhile, the participants in G2 demonstrated only a slight decrease in the amount of carbon emission. The disparities in the amount of carbon emission may be due to the travel distance and the choice of transport for travel. Prior to the implementation of intervention programmes, these participants may have travelled further by car; thus, resulting in higher amount of carbon emission. The implementation of intervention programmes may subsequently prompt these participants to be more aware of their daily travel and commit to sustainable travel behaviour by reducing their travel distance or opting for bus to travel instead.

In conclusion, the obtained results demonstrated the effectiveness of TBM programme in influencing the daily travel behaviour of adolescents. The combination of various intervention strategies, such as motivation, journey planning and incentives, is likely to prompt the adolescents to switch their choice of transport for travel to sustainable transports in the attempt to reduce their carbon footprint. Given the small scale of the case study involving the adolescents in Penang Island across eight selected schools, there is a need to expand the scope of study for future research in terms of the targeted participants. Evidently, the planning and implementation of effective transport policies require a comprehensive grasp on the daily travel pattern of the public to induce sustainable travel behaviour. Moreover, Song et al. (2012) stated that most studies only focused on weekdays travel patterns while the policy-making and the infrastructure planner have been ignoring the weekends travel demands due to its instability. Therefore, it is important to understand the individualistic behaviour and patterns of travel based the variations between weekends' and weekdays' activity-travel characteristics for more effective approaches and management of urban transportation issues especially in developing countries towards low-carbon mobility

References

Agarwal A (2004) A comparison of weekend and weekday travel behavior characteristics in urban areas. Dissertations, University of South Florida. http://scholarcommons.usf.edu/etd/936. Accessed 16 June 2018

Hunt JD, McMillan P, Stefan K, Atkins D (2005) Nature of weekend travel by urban households. In: Annual conference of the transportation association of Canada, Calgary, Alberta. http://conf.tac-atc.ca/english/resourcecentre/readingroom/conference/conf2005/docs/s9/hunt-2.pdf. Accessed 10 April 2018

Jang YJ, Lee SI (2010) An impact analysis of the relationship between the leisure environment at people's places of residence in Seoul and their leisure travel on weekends. J. Korea Plan. Assoc 45:85–100

Koo JK, Kim SS (2004) A study on the improvement of the weekend traffic system affected by implementing the five-day work week. Busan Development Institute, Busan

McDonald NC (2005) Children's travel: patterns and influences. University of California Transportation Center

Song MS, Kim SS, Chung JH (2012) A study on weekend travel patterns by individual characteristics in the Seoul metropolitan area. In: International conference on transport, environment, and civil engineering (ICTECE 2012). http://psrcentre.org/images/extraimages/23%20812553.pdf. Accessed 16 April 2018

Yai T, Yamada H, Okamoto N (1995) Nationwide recreation travel survey in Japan: outline and modeling applicability. Transp Res Rec 1493:29–38. http://onlinepubs.trb.org/Onlinepubs/trr/1995/1493/1493-004.pdf. Accessed 10 June 2017

Zhong M, Hunt JD, Lu X (2008) Studying differences of household weekday and weekend activities: a duration perspective. Transp Res Rec 2054:28–36. https://doi.org/10.3141/2054-04

Printed in the United States
By Bookmasters